Mine Surveying

Sateyesh Singh

RANDOM PUBLICATIONS
NEW DELHI (INDIA)

Mine Surveying

ISBN 978-93-5111-399-7

Published in 2014 in India by

RANDOM PUBLICATIONS

4376-A/4B, Gali Murari Lal, Ansari Road
New Delhi-110 002
Phone : +91-11-43580356, +91-11-23289044
e-mail: randomexports@gmail.com, sales@randompublications.com, info@randompublications.com

Reprinted 2025

Type Setting by : Keystoneprintads, Delhi-110051
Printed at : Replika Press Pvt. Ltd.

Preface

Mine surveying is a branch of mining science and technology. It includes all measurements, calculations and mapping which serve the purpose of ascertaining and documenting information at all stages from prospecting to exploitation and utilizing mineral deposits both by surface and underground working.

Mine surveying is an inseparable part of all stages of mining operations. The work is done by a mine surveying service that is part of the mining enterprise or by an organization in charge of exploring the deposit and the planning and construction of the mining enterprise. For a detailed exploration of mineral deposits the tasks of the mine surveying service include construction of a reference grid, surveying the earth's surface, transferring the plan for the location of exploratory workings of the actual site, and surveying all exploratory workings, as well as natural and man-made rock exposures. Working with geologists, mine surveyors use the surveys to compile the diagrams for mining documentation; the diagrams reflect the volume of exploratory work that has been done, the condition of the earth's surface, the shape and dimensions of bodies of minerals, the quality of the minerals, and the properties of enclosing rock.

Mine surveyors work in the mining science and technology fields, and their key focus is on measurements, calculations and mapping for all stages of the mining process. Mine surveyors are involved in the accurate measurement of areas and volumes mined and the precise representation of surface and underground mining on mine plans. They advise mining engineers about geological areas that can be profitably mined.

The survey is the art of making accurate observations and measurement in the field in order to determine the accurate positions of points on the earth's surface. Mine Surveying is a specialized branch of survey discipline. It is an interesting combination of underground and office work. Qualified surveyors are vital to any mining organization. It is a very responsible job as accuracy of survey details is extremely important for planning purposes.

This book will be a useful asset for the students, researchers and teachers who want to have basic knowledge and other useful information in the area of this field.

I thank all members of my team who have helped in the preparation of the book. My special thanks go to "Random Publications" who have published the book.

– ***Sateyesh Singh***

Contents

Preface .... *v-vi*

1. **Introduction** **1**

Engineering and Mining Surveying 3
Compiling Mine Plans 4
Survey Standards and Procedures 5
Mine Workings Plan 8
Standards for the Preparation of Mine Spatial Information 17

2. **Mining Engineering** **19**

Landfill Mining 39
Mine Development and Production 41
Discipline Mining Engineering 45
Environmental Impact of Mining 49

3. **Types of Mining and Techniques** **53**

Surface Mining 53
Open-pit Mining 58
Mountaintop Removal Mining 60
Underground Mining (Hard Rock) 67
Underground Mining (Soft Rock) 75
Drift Mining 76
Shaft Mining 77
Slope Mining 80
Retreat Mining 80
Longwall Mining 81
Hydraulic Mining 86
Copper Extraction Techniques 93

4. **Mining Equipment** **103**

The Modern Art of Lubricating Mining Equipment 103
Steam Shove 109
Power Shove 116
Excavator 119

Dragline Excavator 123
Bucket-wheel Excavator 130
Dredging 134
Dynamite 144
Blasting Machine 147
Hoist (Mining) 148
Beam Engine 151
Drilling Rig 153
Wellhead **170**

5. Mining Processing 173

Mining–Open Pit and Underground 174
Mineral Processing 175
Froth Flotation 183
Tromme 191
Gold Panning 192
Rocker Box 194
Drywasher 196
Extractive Metallurgy 197
Pyrometallurgy 199
Cupellation 201
Hydrometallurgy 204
Leaching (Chemistry) 207
Amalgam (Chemistry) 209

6. Land Surveying 213

Overview of Licensing Board 4
Industry-specific Features 219
Relevance and Independence 221
Future Arrangements for Board 232
Review Assessment: Other Issues 237

7. Coal Mining Geospatial Data Standards 239

Introduction 239
Fitness for Work Policies in Coal Mining 251

8. Mine Planning 260

Planning Related to Mining Size 262
Planning Related to Physical Properties 265
Planning the Organization and Required Equipment 2280
Mine Planning and Design (MPD) 283
Mine Design Methods 289
Ground Support And Reinforcement (GSR) 291

***Index* 301**

1

Introduction

The branch of mining science and engineering that uses measurements made in the field, and also subsequent geometric constructions, as a basis for the study of the structure of a deposit, the shape and dimensions of mineral bodies in the interior and the location in them of components that are useful and harmful (for mining technology), the properties of enclosing rock, the spatial arrangement of mine workings, and the processes of stresses in rock and the earth's surface in connection with mining work. It also reflects the dynamics of the production process of a mining enterprise. Mine surveying work is done by means of mine surveying instruments.

Data from mine surveying are synthesized in mine documentation, which is a set of diagrams produced by geometric projection. Mine surveying is a composite science that is closely related to many scientific disciplines: mathematics, engineering, physics, astronomical geodesy, geology and mineralogy, geological exploration, the technology of exploitation of deposits, and construction. Mine surveying is an inseparable part of all stages of mining operations. The work is done by a mine surveying service that is part of the mining enterprise or by an organization in charge of exploring the deposit and planning and construction of the mining enterprise.

For detailed exploration of mineral deposits the tasks of the mine surveying service include construction of a reference grid, surveying the earth's surface, transferring the plan for the location of exploratory workings to the actual site, and surveying all exploratory workings, as well as natural and man-made rock exposures. Working with geologists, mine surveyors use the surveys to compile the diagrams for mining documentation; the diagrams reflect the volume of exploratory work that has been done, the condition of the earth's surface, the shape and dimensions of bodies of minerals, the quality of the minerals, and the properties of enclosing rock. Mine surveyors also take part in estimation of geological reserves.

During the planning of mining enterprises, mine surveyors take part in planning and surveying work, planning the boundaries of mining enterprises and systems for working the deposit, locating projected buildings and structures in the area of the mineral deposit, determining safety measures to protect structures against the harmful influence of mining development, and compiling timetables for the development of mining work. In addition, they check planned ratios of the geometric elements of the master plan for the surface and make calculations of available reserves. During the construction of mining enterprises the mine surveying service constructs a grid of reference points, transfers the geometric elements of the construction plan to the actual site, checks on fulfillment of the planned ratio of geometric elements during installation of hoisting devices and reinforcing of mineshafts, assigns the direction of mine workings, makes execution surveys, and prepares the mining documentation necessary for exploitation of the deposit.

During the exploitation of deposits the mine surveying service records the mining enterprise's production process and compiles the drawings for mine documentation. As the mine workings progress, the service establishes with greater accuracy the conditions of bedding of the deposit and the shape of the mineral body, the quality of the mineral, and the properties of the enclosing rock. In addition, it compiles diagrams that reflect the structure of the deposit, the shape of the mineral bodies, the quality of the mineral and distribution of useful components, the properties of enclosing rock, the dynamics of the process of rock shifting, and other phenomena that must be known in order to improve the technological procedures for exploitation of the deposit and to plan the development of mining work.

The direction of mine workings is set, steps are taken to ensure safety near danger zones and checks are instituted to see that such measures are taken, observations are made on the shifting of rock and on phenomena of rock pressure, measures are developed to protect buildings, structures, natural objects, and mine workings against harmful influence from mine development, and records are kept of the movement of industrial reserves, losses, and depletion of the mineral. When mining enterprises are shut down permanently or temporarily, the mine surveying service determines to what extent the mineral has been extracted and updates the mine documentation diagrams to the time of shutdown.

Mine surveying originated simultaneously with mining. Problems of measurements in mineshafts were being solved even in ancient times. In Egypt, as early as the 16th to 14th centuries B.C., reduced scale representations of mining excavations were constructed on plane surfaces. Hero of Alexandria (presumably first century A.D.) was the first to describe the method of underground surveying and orientation. The German scholar G. Agricola (16th century) gave a brief summary of methods of accomplishing tasks of mine surveying. In Russia, mine surveying finally took shape under Peter I as a special department of mining enterprises. At that time the specialists who worked on spatial geometric measurements related to mining were called mining geometers. Their experience was first summarized by M. V. Lomonosov in his work *The First Foundations of Metallurgy* (1763).

ENGINEERING AND MINING SURVEYING

Engineering and Mining Surveyors provide their professional advice and skills for the building and development of various aspects in our modern life. This includes the construction of our homes to the development of our towns and cities and the links between them.

These surveyors are usually the first to be called upon for any development, large or small to provide the technical data required by other professionals to undertake their roles of planning and design. This usually involves these surveyors taking measurements of the natural environment using various forms of equipment including using satellite measuring equipment, laser scanners, as well as specialised surveying equipment such as Total Stations and Levels. Using this equipment and the measurements taken they will provide accurate models and data of the existing form in various formats.

But their role in our modern world does not stop here. Engineering and Mining Surveyors are the ones who help bring the developments to life, using the plans provided by other professionals, they provide the control and ongoing supervision of works by construction crews. Their measurements and survey marks are relied upon to ensure the development is constructed as per the design as well as any redesigns or amendments that may be required along the way.

Once construction is completed, Engineering and Mining Surveyors are asked to come back and take measurements of the completed development to make sure the development has been built as per the design within specification as well as record as-built information for the protection of many of the services we rely on in modern life.In some regions Mining Surveyors are registered or authorised as their work is relied upon for the safety of present and future workers within the mine workings as well as the ongoing monitoring of the mine site during and after mining. Engineering and Mining Surveyors may also extend their skills and undertake various other aspects in the development process which are undertaken by other professionals including planning, design and project management.

COMPILING MINE PLANS

The mining operation may operate on a local survey grid/datum however all Mine Workings Plan and Mine Spatial Information should be compiled on the Map Grid of Australia 1994 (MGA94) based on the Geocentric Datum of Australia 1994 (GDA94) values.

The Mine Workings Plan should be sectionalised into sheets that best fit individual mines. In determining the best fit for the sheet layout the Mine surveyor should be mindful of the need to provide for expansion of the mine and adjoining mine workings.

Where old workings exist it should be assumed, for the purpose of marking the Mine Workings Plan, that they constitute a danger until proven otherwise. All plans should be regarded with suspicion until their accuracy has been verified and validated. All reasonable effort should be made to obtain all existing information about old workings which should be recorded on the Mine Workings Plan.

Symbols

The technical symbols, sign conventions and definitions for strata to be shown on the Mine Workings Plan and the Mine Spatial Information should be in accordance with:

- *Australian Standard for Mine Plans – Preparation and Symbols (AS-4368),* and
- *Australian Standard for Geological Symbols (AS-2916).*

If a symbol is not provided for in AS4368 or AS2916 the Mine surveyor may create a suitable symbol to be shown in the legend.

Functions of a Mine surveyor

The functions of the Mine surveyor include the following:

- To competently develop mines plans Bring to the attention of the mine operator:
  - Any variation in mine workings from a Mine Workings Plan of which the surveyor becomes aware, or
  - Any interference with or obstruction to the surveyors' performance or functions, or
  - Any doubt about the accuracy of any plans, *Conversion to Map Grid of Australia 1994 (MGA94)*

When new sheets of the Mine Workings Plan are being prepared and the workings of any part of the mine should be completed, sealed or otherwise become inaccessible, it is acceptable to draw an outline of such workings and to endorse the new Mine Workings Plan and to refer to the previously prepared Mine Workings Plans for detail. Any previously prepared Mine Workings Plan may also be a historical set of mine drawings that are accessible and in good condition. The relationship of the origin and height datum of any set of drawings to the current mine grid and datum or to MGA94 and AHD should be established. Should the workings become active, or used for access in an area previously shown in outline, the survey of the workings should be transferred in full to the new Mine Workings Plan. Nothing should prevent the Mine surveyor from transferring inactive mine workings in full to the new Mine Workings Plan.

Inspection

The mine operator must keep the mine survey plan available for inspection under the Act. It must also be readily accessible on request to workers at the mine.

SURVEY STANDARDS AND PROCEDURES

Origin of Co-ordinates

All surveys should originate from the Mine baseline or any other mark included in the state survey control network having a standard of accuracy consistent with that of the Mine Baseline. The survey of the Mine Baseline should be planned and surveyed to Class "B" standards of accuracy as defined in ICSM SP1.

Where required, at each end of the Mine Baseline there should be a reference mark of durable nature connected by bearing and distance to the baseline permanent mark. If the position or co-ordinate values of the Mine Baseline change, this information should be reported in accordance with any regulations.

Where required, a hard copy plan and an electronic record with digital survey data should be forwarded to any regulatory body. For an existing mine, the current baseline marks should be retained provided they are of a durable nature.

Underground Baseline

Each underground mine should establish underground baselines, in suitable positions in the underground workings and as long as practicable. The terminal marks should be stable and durable. Underground Baseline details should be recorded on the Mine Workings Plan.

CONTROL SURVEYS AND SUBSIDIARY SURVEYS

Where no prescribed accuracy is defined, each Control Survey should achieve a standard of accuracy as prescribed in ICSM, SP1 to Class D or better. When calculating compliance to ICSM SP1 Class D using the formula r = c (d + 0.2) this direction specifically defines d as the following: d = distance to any station in km, with a minimum value of 1(km). Marking: each control survey station should be adequately referenced and substantially marked. As far as practicable the marks should be placed in a position least likely to be disturbed.

Subsidiary surveys should be employed by the Mine surveyor where necessary to within an accuracy of 2m of position.

Correlation of Surface and Underground Surveys

Correlation between surface and underground surveys should be consistent with a Class D survey as prescribed in ICSM SP1. In correlation of surface and underground surveys where methods other than direct traverse are employed for azimuth or co-ordinate transfer, the surveys should be shown on a plan separate from the Mine

Workings Plan and should disclose the special survey methods employed. The plan should be retained at the operation and the Mine surveyor should certify that the survey shown on the plan is accurate.

Survey Records and Supply of Survey Information

Systematic and reasonable care should be taken to ensure the safe preservation of all survey records. Records should be kept at the survey office for each of the following purposes:

- Control surveys,
- Subsidiary surveys,
- Surface movement and subsidence surveys.

Survey Records should be maintained either manually (in a field book or other stable material) or electronically. Where a machine or other device is required to access the stored data the Mine surveyor should ensure the data is regularly updated to a media and format that is currently available. Survey Records should be permanently recorded and maintained so that they can be utilised for future surveys

- All survey books should be maintained in good order and should have the following description clearly and permanently marked on the cover and inside title page:
  - The name of the mine, and
  - For underground mines, the level name and catalogue number to which the book refers, and
  - Consecutive index number.
- The following procedures should be adopted for entries into survey books
  - All survey observations and measurements should be recorded at the time of survey;

- In the event of alteration of a mistake there should be no erasure
- The erroneous entry should be struck through and the correction written above
- The datum line of the survey and the azimuth adopted should be clearly indicated
- Lengths should be entered at the time they are measured and where necessary corrections should be clearly indicated
- Bearing and distance from reference marks should be clearly shown
- Reference marks and bench marks placed by the surveyor should be noted with the positions and descriptions shown by a sketch in the appropriate book
- Lines remeasured should be so specified and original distances and bearings shown, and
- The mine surveyor should sign and date the field book showing the work was performed by him or under his supervision.

Where surveys are recorded in electronic form the information recorded should be consistent with that required for survey books. A complete and separate duplicate of such records should be preserved on paper or disc or other permanent electronic medium. Where a machine or other device is required to access the stored data the mine surveyor should ensure the data is regularly updated to a media and format that is currently available . The mine surveyor should make available any records (or copies) to the regulatory body when requested. Upon Discontinuance of a mine all Survey Records relevant to the preparation of the Mine Workings Plan should be prepared by the Mine surveyor for submission to the regulatory body for retention.

Requirements before Workings Become Inaccessible

Before any part of the workings of a mine becomes inaccessible the position of the workings should be established from a control or subsidiary survey (where reasonably practical and safe to do so). Where inaccessible workings are unable to be surveyed; the Mine surveyor should take all reasonably practicable steps to assist completion of the Mine Workings Plan.

Surface Movement and Subsidence

Surveys should be undertaken to record surface movement, including subsidence induced by mining. These surveys should be carried out under the supervision of, and certified by, the Mine surveyor. A copy of all surface movement and /or subsidence survey data including field notes should be retained by the mine operator.

Survey of Boreholes

Boreholes should be identified by unique name or number and should be shown on the Mine Workings Plan. Collar and/or commencement locations of Boreholes, should be established from a control or subsidiary survey. The following information should be utilised, when available, from direct survey or other source to define the position of a borehole and suitably endorsed (for example a driller's log or geophysical survey):

- Total length (or depth);
- Inclination (or declination); and
- Plan projection (bearing or azimuth) when other than near vertical.

MINE WORKINGS PLAN

PREPARING THE MINE WORKINGS PLAN

The Mine Workings Plan should be prepared by or under the supervision of the Mine surveyor.

Updating the Mine Workings Plan

Except where the mine is considered discontinued or abandoned, the WHS Regulations require the Mine Workings Plan to be updated:

- When there has been a significant mine modification; and
- At least every 3 months in relation to the parts of the plan identify points of access and egress and refuges; and
- Otherwise at least annually.

Composition of the Mine Working Plan

In the case of an underground mine the Mine Workings Plan should comprise sheets for each Seam or Level, Cross Sections and Longitudinal Sections, and other layout data for the whole of the Mine as required by this code.

In the case of an open cut mine, the Mine Workings Plan should be compiled on the Open Cut Void Sheets and Cross Sections.

Presentation and Archiving–hard copy

A hard copy of the Mine Workings Plan should be produced on stable and durable material. The hard copy will comprise all sheets on which workings have occurred.

Level, Seam and Open Cut Void Sheets

The underground mine workings or open cut voids should be mapped on Level Sheets, Seam Sheets or Open Cut Void Sheets. The Level Sheets, Seam Sheets and Void Sheets should be compiled at a scale that best reflects the mine workings with sufficient clarity and number of sheets to cover the extent of the workings and necessary adjacent workings.

Information to be Shown on the Mine Workings Plan

The WHS Regulations require that the mine plan show:

- The workings of the mine, including disused workings and incorrectly located bores
- The ventilation system, including all ventilation fans
- The location of electrical installations associated with the distribution of electricity at the mine
- The location of telephones and other fixed plant associated with the radio and telecommunications systems
- Water dams and tailings dams
- Natural features surrounding the mine
- Places where hydrocarbons or explosives are stored, and
- Points of access and egress, including emergency exits.

General information for both underground and open cut mines:

- Adjacent mine workings - an outline of all workings in any direction within 100 metres of the current mine being worked.
- Barriers and restricted zones.
- Borehole locations:
  - Where it is not practical to show all Boreholes on the Level Workings sheet, a separate "Borehole Plan", which should become part of the Mine Workings Plan, should be produced as an overlay.
  - A reference on the Level Workings Sheet is required to indicate the existence of any Borehole Plan sheet.
  - All Boreholes should be identified as to type and name.
  - Any borehole removed by the mining process or is in the void is not required to be shown. Remnant stubs of these holes that may present a hazard to future mining operations should be shown.
- Cadastral Parcels.
- MGA Grid Lines.
- Mining Approvals/ Acceptances (if applicable from State prescribed legislation).
- Mining lease boundaries.
- Potential sources of Inrush to be shown (outlined in green).
- State regulated dams
- Survey control stations including bench marks.
- Surveyed geological features - all known outcrops, subcrops, faults, dykes, cinder belts, and other significant dislocations, as surveyed.
- Mine operation boundaries.

Underground mines - Level Sheets:

- Development.
- Mine underground access mine shafts, rises, winzes, adits, and declines.

- Stopes.
- Inferred block/sublevel cave zone of influence
- Date lines (dashed) indicating the extent of workings for each survey reporting period along with the date and initials of the Mine surveyor.
- Working level floor heights.
- Level names.
- All major ventilation devices.

Underground mines - Seam Sheets:

- Development.
- Mine underground access - drifts, mine shafts and adits.
- Extracted Areas.
- Date lines (dashed) indicating the extent of workings for each survey reporting period along with the date and initials of the Mine surveyor.
- Working level contours.
- Panel names.
- All major ventilation devices

Open cut mines (Open Cut Void Plans):

- Void contours at five (5) metre vertical intervals.
- Void detail line (breaklines)

Sheet heading Information:

- The name of the mine.
- The Mine Workings Plan Catalogue Number, where available
- The name of the level/seam being represented.
- The number of the sheet and the number of sheets that make up the mine operation.

Map surround area information:

- The reduction ratio and a graphical (bar) scale together with a statement that all measurements are in metres.
- A north point indicating grid north.
- A schedule of symbols used on the particular sheet.
- A sheet index showing (a vertical level index may be necessary):
  - all the sheets necessary to cover the mine operations boundary; and
  - the number of each sheet; and
  - the particular sheet shown by a heavy outline; and
  - the outline of mine workings.
- A schedule of endorsements made by the Mine surveyor,
- A statement or schedule certifying the accuracy of the sheet, including:
  - name and signature of the certifying Mine surveyor; and
  - certifying surveyor's identification/authorisation/certificate number; and

  - date of the certification.
- The origin of levels and the grid bearing and terminal survey stations of the mine baseline or survey control marks used for surveys within the sheet.
- A schedule or schedules of survey marks and bench marks containing their identification, coordinates and height.
- For underground mines, a schedule of shafts, rises, winzes, boreholes, with coordinates and heights of the commencing and finishing locations.
- A schedule of boreholes showing names, coordinates and heights of the collar.

Additional Information

The Mine surveyor should show on the Mine Workings Plan any additional information required by the regulator . Any additional information discovered by or indicated to the Mine surveyor which may present a hazard either to the mine, adjacent mines or persons in those mines should be recorded on the Mine Workings Plan.

Doubtful Information

Where any information shown on the Mine Workings Plan is considered to be in doubt, the Mine surveyor should identify and endorse this information.

Old workings/surveys

Where old surveys or old workings are converted to the current mine Grid/Datum or GDA 94/MGA94 such conversion should be endorsed by the Mine surveyor on the Mine Workings Plan.

Certification

Unless prescribed by State legislation, the following certification should be utilised.

- The hard copy of the Mine Workings Plan should be certified (signing and dating) by the Mine surveyor in the certification of accuracy schedule; after the most recent Reporting Period has been charted.
- The Mine surveyor by signing and dating the certification of accuracy schedule declares:
  - the Mine Workings Plan has been prepared ; and
  - the surveys shown on the Mine Workings Plan have been completed to an accuracy .

Certification history

The Mine Workings Plan should have a schedule of certification of accuracy schedule, showing the certification details for each plan update.

Catalogue number

Any regulatory body cataloguing system in force should continue for each mine. This unique number is to be clearly identified on the Mine Workings Plan or any copy thereof.

MINE SPATIAL INFORMATION

This section refers to the preparation, composition and supply of the Mine Spatial Information as a digital file.

Preparation

The Mine Spatial Information should be prepared by or under the supervision of the Mine surveyor. The Mine Spatial Information should be prepared to the standards required by this code.

Composition

The Mine Spatial Information should be digital files of the whole of the mine. Digital files should be in 3D format and should be compatible with the regulatory body software. The following Mine Spatial Information is the minimum themes to be supplied, as appropriate for each mine:-

- For underground mines – Mine Workings (wall, roof and floor), Extraction areas.
- For open cut mines – Mine Workings Outline, Open Cut Void Breaklines and Open Cut Void Points

CLOSING PLANS

Discontinuance

Where a mine has been discontinued the mine operator should have the Mine Workings Plan charted to the date of discontinuance.

Charting for Discontinuance

These procedures should be followed in the case of the discontinuance of a mine:

- The Mine Workings Plan should be charted, dated and signed by the Mine surveyor to the date of discontinuance. The Mine surveyor should place a broken line around the extent of the workings, which should be dated and initialled.
- The note "Charted to date of Discontinuance" is to be shown with the date and the Mine surveyor's signature.

Charting for Abandonment

These procedures should be followed in the case of the abandonment of a mine:

- The Mine Workings Plan should be charted, dated and signed by the Mine surveyor to the date of abandonment. The Mine surveyor should place a broken line around the extent of the workings, which should be dated and initialled.
- The note "Charted to date of Abandonment" is to be shown with the date and the Mine surveyor's signature.

Cessation of duties of the Mine surveyor

Upon permanent cessation of duties of the nominated Mine surveyor, the Mine Workings Plan should be charted, dated and signed by the Mine surveyor. The Mine surveyor should show the date of the workings at the time of cessation in a similar manner to that of the normal survey reporting period. The note 'Charted to date of Cessation of duties of the Mine surveyor' is to be shown with the date and the Mine surveyor's signature. The Mine surveyor by signing and dating declares (for the period from the last reporting period until the date of cessation of duty):

- The Mine Workings Plan has been prepared.
- The surveys shown on the Mine Workings Plan have been completed to an accuracy prescribed in this code.

Replacement of Business Operator

When the mine operator ceases to operate the mine they should update the mine workings plan and provide the updated plan to the incoming operator and/or mine holder.

SUPPLY OF MINE WORKINGS PLAN AND DIGITAL DATA

Supply Period

Where required the mine surveyor should supply to the regulator, in the specified time, copies of mine working plans and digital data.

Supply on Discontinuance and Abandonment

Where a mine has been discontinued, the Mine Workings Plan should, after charting, be supplied to regulator. The Mine Spatial Information should be supplied on a CD-R (Compact Disc – Read only) or equivalent. The following information will be recorded on the label of the CD-R (or equivalent) with a permanent marking pen:

- The name of the mine operation.
- The Mine Workings Plan catalogue number for the associated Mine Workings Plan.
- The name and signature of the certifying Mine surveyor.
- The certifying Mine surveyor's registration Number.
- The date of certification.

Upon Discontinuance of a mine, all survey records relevant to the preparation of the Mine Workings Plan, should be prepared by the Mine surveyor for submission to the regulatory body.

SUPPLY ON CESSATION OF DUTIES OF A MINE SURVEYOR

At the cessation of duties after certification has taken place, the Mine surveyor should produce a hard copy of the Mine Workings Plan and a copy of the Mine Spatial Information electronically for retention at the office for the mine.

The outgoing Mine surveyor, with the consent of the person conducting the business, should take a copy of the Mine Workings Plan and Mine Spatial Information for their own records. The incoming Mine surveyor should make a copy of the Mine Workings Plan and Mine Spatial Information for a record of commencement of work.

Certification

The Mine surveyor, by signing and dating the CD-R, declares that the Mine Spatial Information contained on the CD-R, were used in the preparation of the Mine Workings Plan.

OTHER PLANS REQUIRED

Unless prescribed by State legislation, the following plans should be utilised for mining operations. These plans are a minimum requirement and does not prevent the Mine surveyor or the mining operation from generating further plans for purposes of emergency response or ongoing mine management.

Combining two (2) or more plans

The Mine surveyor can combine two (2) or more of the following plans provided legibility of the combined plan is retained.

Endorsing combined plans

Combined plans should be prepared and should be certified and endorsed by the Mine surveyor and Mine Manager (or equivalent statutory official).

Subsidence Management Plans

These plans may be required by Mining Lease conditions and are prepared in accordance with the regulatory body guideline. Plans prepared under the guideline should be prepared in accordance with the standards required by this code. All such plans will be certified by the Mine surveyor and suitably endorsed where necessary. Regulatory bodies may publish guidance notes for applications under legislation and the subsidence management plan process and these should be referred to

Ventilation Plan (underground operations)

The ventilation plan should be compiled at a scale that best reflects the mine workings and ventilation appliances. The plan should show ventilation walls, trapdoors, regulators, ventilation doors, the direction and nature of ventilation, auxiliary fans, booster fans and air reading stations (with air quality and quantity), in relation to the mine workings. The plan will also show the locations of fixed gas monitoring plant. A legend will be shown on the plan depicting symbols used together with a graphical representation of the scale. This plan should be updated monthly.

Fire Fighting Plan (Underground Mines)

A Fire Fighting Plan is to be prepared for all underground parts of an operation. The Fire Fighting Plan should be compiled generally at a scale not less than 1:10,000. Fire fighting appliances will be symbolised in accordance with AS 4368. The plan should show positions in which pipe mains, hydrants, isolation valves, fire substations and fire depots are situated. This plan will also show the position of stoppings, trapdoors, regulators, seals, prepared seal sites, overcasts, air crossings, ventilation doors, belt conveyors, main electric supply cables, fixed electrical apparatus, high voltage cables and switchgear, telephones and the direction of ventilation in the roadways. The plan is to be updated at no greater than three month intervals.

Escape and Rescue Plan (underground operations)

The Escape and Rescue Plan should be compiled generally at a scale that best reflects the mine workings and displayed appliances. The plan will include all workings of the mine to enable the self escape and/or rescue of personnel from anywhere in the mine. Symbols used will be in accordance with AS 4368 and AS2916. This plan should show main roads, the means of egress from each part of the mine to the surface, all self-escape systems and all underground telephone stations, in relation to the mine workings.

This plan should also show main access roads, the means of egress from each part of the mine to the surface, all self-escape systems, refuge chambers, the positions of first aid stations, underground telephone stations, belt conveyors, any gas drainage mains and branches, known falls, accumulations of water, stowage identified as non-passable. The plan should also show any information from the fire fighting plan and the ventilation plan or any additional information identified as necessary for the dealing with an emergency at the operation.

The plan should also show any information including ventilation or any additional information identified as necessary for the dealing with an emergency at the operation. The plan should be displayed, reviewed and updated every three months. The regulation requires review, and if necessary revision, when there has been a significant mine modification, at least every three months in relation to the identification of points of access, egress and refuges, and otherwise annually.

Surface Plan

A surface plan should be kept at the mine. The Surface plan should be of a scale to best represent the surface features of the mine and should cover the areas where mining operations have been or are being carried out and construction zones, as required by any regulation. Topography may be shown if it adds to the interpretation of the plan. Compilation should be generally and the plan may be maintained as spatial information.

In the case of an underground mine, the Surface Plan should show all streets, roads, reservoirs, swamps, water bodies, unconsolidated surface deposits, railways, main pipelines and any other feature whether of the same or of a different kind which, if disturbed by mining operations, is likely to cause damage to or danger in the mine. In the case of an open cut mine, the Surface Plan should show all streets, roads, reservoirs, swamps, water bodies and any other permanent feature whether of the same or of a different kind which, if disturbed by mining operations, is likely to cause damage to or danger in the open cut workings. This plan should be updated as often as it becomes necessary.

Site Services Plan (underground and open cut mines)

The site services plan should be at a suitable scale. Symbols used will be in accordance with AS 4368. Compilation should be generally and the plan may be maintained as spatial information. This plan should show only surface improvements and infrastructure that are part of the mine, including but not limited to:

- Buildings including administration, bathhouse and workshops, mine access points (portals, shafts, ramps);
- Mine access roads including parking areas;
- Processing plants and stockpiles;
- Power reticulation including sub stations;
- Water reticulation;
- Compressors and compressed air reticulation;
- Gas reticulation;
- Sewer mains;
- Treatment plants and transpiration areas;
- Telephone and other communications lines;
- Magazines including buffer zones;
- Fire fighting equipment including fire stations, hydrants, extinguishers, depots;
- Hazardous materials locations;
- Dams and surface drainage;

- Boreholes which can be used for gas testing;
- Underground/buried services and cable locations;
- Any information which may assist in the case of an emergency;

Topography may be shown if it adds to the clarity and understanding of the plan. This plan should be updated as often as it becomes necessary.

TAILINGS DAMS/WASTE DUMP AREAS

These plans should be prepared in accordance with any relevant guidelines. The guidance contained in this code should be used when developing tailings or waste dump plans. The plan should be certified and suitably endorsed by a Mine surveyor.

Emplacement Areas

These plans should be prepared in accordance with any relevant guidelines. The guidance contained in this code should be used when developing emplacement plans. The plan should be certified and suitably endorsed by the Mine surveyor.

Plan standards-general

Any plan (including, where appropriate, plans subject to this code) required to be drafted by the Mine surveyor for purposes of the mine, should be prepared in accordance with the relevant Australian Standard in particular, but not limited to, AS 4368, AS 1100 and AS 2916.

Declaration

A plan referred to in Section 8 of this code should have an appropriate area on the plan allocated for certification by the mine manager (or equivalent statutory position) and the Mine surveyor. Certification should indicate the origin of the information and that the information shown on the plan is truly represented.

STANDARDS FOR THE PREPARATION OF MINE SPATIAL INFORMATION

This section:

- Provides for the construction and preparation of the Mine Spatial Information.
- Identifies how each theme should be represented graphically and the attribute information that should be included to describe each featured object in the theme.
- Identifies supporting information that is required may have to be submitted electronically.
- Assist in the preparation of spatial information in support of such requirements for subsidence management plans, mining operations

plans, annual environmental management reports and periodic geological reporting.

Attribute Values

Where possible standard attribute values should be used. This will allow other users of the themes to develop standard enquiries through their GIS software. Values for all attributes are required to be attached for all objects in the theme. All themes should be prepared using the database format. Numbers with decimal values should be stored as floating number. All other numbers should be stored as integers or characters.

Date

Date attributes should be in database format 'character' and be in the form "dd/mm/yyyy" where date of workings are accurately known, general format "month year" (May 2006) or as a time period (1950-2000 or pre 2000) when the charted period cannot be accurately shown.

Metadata Statements

The metadata statement should include information such as version date, custodian name, coordinate system, accuracy, potential limitations for data use, definitions of any codes used within the dataset and any other information about the data that is not evident in the attributes attached to the theme. The completed metadata statement should be disseminated with the dataset that it describes. It may be one document sectionalised for each Mine Spatial Information or a separate document for each theme. Where two or more themes or a group of themes have the same metadata, a single metadata statement may be prepared that clearly lists all themes relating to the statement.

Precision

Spatial themes should be provided in the best available precision.

Preparation of Additional Themes

Additional themes should be consistent with this code.

Preparation of Additional Attributes

Any additional attributes (including formats) should be developed consistent with this code.

2

Mining Engineering

Mining is the extraction of valuable minerals or other geological materials from the earth from an orebody, lode, vein, seam, orreef, which forms the mineralized package of economic interest to the miner. Ores recovered by mining include metals, coal and oil shale, gemstones, limestone, and dimension stone, rock salt and potash,gravel, and clay. Mining is required to obtain any material that cannot be grown through agricultural processes, or createdartificially in a laboratory or factory. Mining in a wider sense includes extraction of any non-renewable resource such as petroleum,natural gas, or even water.

Fig. Surface coal mining

Mining of stone and metal has been done since pre-historic times. Modern mining processes involve prospecting for ore bodies, analysis of the profit potential of a proposed mine, extraction of the desired materials, and final reclamation of the land after the mine is closed.

The nature of mining processes creates a potential negative impact on the environment both during the mining operations and for years after the mine is closed. This impact has led to most of the world's nations adopting regulations to moderate the negative effects of mining operations. Safety has long been a concern as well, and modern practices have improved safety in mines significantly.

HISTORY

Prehistoric mining

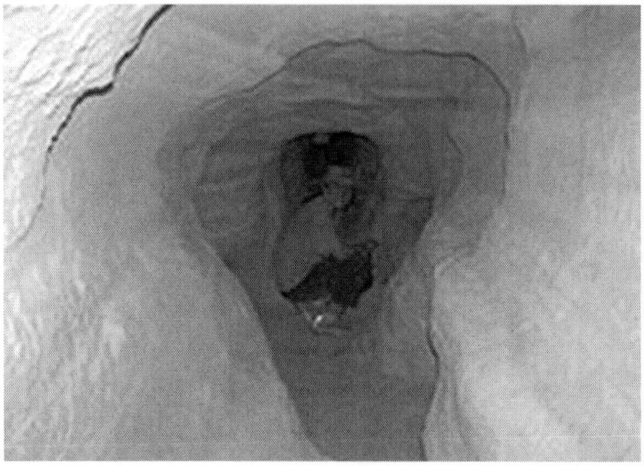

Fig. Chalcolithic copper mine in Timna Valley, Negev Desert, Israel.

Since the beginning of civilization, people have used stone, ceramics and, later, metals found on or close to the Earth's surface. These were used to make early tools and weapons, for example, high quality flint found in northern France and southern Englandwas used to create flint tools. Flint mines have been found in chalk areas where seams of the stone were followed underground by shafts and galleries. The mines at Grimes Graves are especially famous, and like most other flint mines, are Neolithic in origin (ca 4000 BC-ca 3000 BC). Other hard rocks mined or collected for axes included the greenstone of the Langdale axe industrybased in the English Lake District.

The oldest known mine on archaeological record is the "Lion Cave" in Swaziland, which radiocarbon dating shows to be about 43,000 years old. At this site paleolithic humans mined hematite to make the red pigment ochre. Mines of a similar age inHungary are believed to be sites where Neanderthals may have mined flint for weapons and tools.

Ancient Egypt

Ancient Egyptians mined malachite at Maadi. At first, Egyptians used the bright green malachite stones for ornamentations and pottery. Later, between 2613 and 2494 BC, large building projects required expeditions abroad to the area of Wadi Maghara in order "to secure minerals and other resources not available in Egypt itself." Quarries for turquoise and copper were also found at "Wadi Hamamat, Tura, Aswan and various other Nubian sites" on the Sinai Peninsulaand at Timna.

Mining in Egypt occurred in the earliest dynasties. The gold mines of Nubia were among the largest and most extensive of any in Ancient Egypt, and are described by the Greek author Diodorus Siculus. He mentions that fire-setting was one method used to break down the hard rock holding the gold. One of the complexes is shown in one of earliest known maps. They crushed the ore and ground it to a fine powder before washing the powder for the gold dust.

Ancient Greece and Rome

Mining in Europe has a very long history. Examples include the silver mines of Laurium, which helped support the Greek city state of Athens. However, it is the Romans who developed large scale mining methods, especially the use of large volumes of water brought to the minehead by numerous aqueducts. The water was used for a variety of purposes, including using it to remove overburden and rock debris, called hydraulic mining, as well as washingcomminuted or crushed ores, and driving simple machinery.

The Romans used hydraulic mining methods on a large scale to prospect for the veins of ore, especially a now obsolete form of mining known as hushing. It involved building numerous aqueducts to supply water to the minehead where it was stored in large reservoirs and tanks. When a full tank was opened, the wave of water sluiced away theoverburden to expose the bedrock underneath and any gold veins. The rock was then attacked by fire-setting to heat the rock, which would be quenched with a stream of water. The thermal shock cracked the rock, enabling it to be removed, aided by further streams of water from the overhead tanks. They used similar methods to workcassiterite deposits in Cornwall and lead ore in the Pennines.

The methods had been developed by the Romans in Spain in 25 AD to exploit large alluvial gold deposits, the largest site being at Las Medulas, where seven long aqueducts were built to tap local rivers and to sluice the deposits. Spainwas one of the most important mining regions, but all regions of the Roman Empire were exploited. They usedreverse overshot water-wheels for dewatering their deep mines such as those at Rio Tinto. In Great Britain the natives had mined minerals for millennia, but when the Romans came, the scale of the operations changed dramatically.

The Romans needed what Britain possessed, especially gold, silver, tin and lead. Roman techniques were not limited to surface mining. They followed the ore veins underground once opencast mining was no longer feasible. AtDolaucothi they stoped out the veins, and drove adits through barren rock to drain the stopes. The same adits were also used to ventilate the workings, especially important when fire-setting was used. At other parts of the site, they penetrated the water table and dewatered the mines using several kinds of machine, especially reverse overshot water-wheels. These were used extensively in the coppermines at Rio Tinto in Spain, where one sequence comprised 16 such wheels arranged in pairs, and lifting water about 80 feet (24 m). They were worked as treadmills with miners standing on the top slats.

Many examples of such devices have been found in old Roman mines and some examples are now preserved in the British Museum and theNational Museum of Wales.

Medieval Europe

Fig. Gallery, 12th to 13th century, Germany

Mining as an industry underwent dramatic changes in medieval Europe. The mining industry in the early Middle Ages was mainly focused on the extraction of copper and iron. Other precious metals were also used mainly for gilding or coinage. Initially, many metals were obtained through open-pit mining, and ore was primarily extracted from shallow depths, rather than though the digging of deep mine shafts. Around the 14th century, the demand for weapons, armour, stirrups, and horseshoes greatly increased the demand for iron. Medieval knights for example were often laden with up to 100 pounds of plate or chain link armour in addition to swords, lances and other weapons. The overwhelming dependency on iron for military purposes helped to spur increased iron production and extraction processes.

The silver crisis of 1465 occurred when the mines had all reached depths at which the shafts could no longer be pumped dry with the available technology. Although the increased use of bank notes and the use of credit during this period did decrease the dependence and value of precious metals, these forms of currency still remained vital to the story of medieval mining.

In the mid-sixteenth century the great attack on mineral deposits spread from central Europe to England. England had iron, zinc, copper, lead, and tin ores. On the continent all mineral deposits belonged to the crown, and this regalian right was stoutly maintained; but in England it was pared down to gold and silver (of which there was virtually none) by a judicial decision of 1568 and a law of 1688. Landlords therefore owned the base metals and coal under their estates and had a strong inducement to extract them or to lease the deposits and collect royalties from mine operators. English, German, and Dutch capital combined to finance extraction and refining. Hundreds of German technicians and skilled workers were brought over; in 1642 a colony of 4000 foreigners was mining and smelting copper at Keswick in the northwestern mountains.

Use of water power in the form of water mills was extensive; they were employed in crushing ore, raising ore from shafts and ventilating galleries by powering giant bellows. Black powder was first used in mining in Selmecbánya, Kingdom of Hungary in 1627. Black powder allowed blasting of rock and earth to loosen and reveal ore veins. Blasting was much faster than fire-setting and allowed the mining of previously impenetrable metals and ores. In 1762, the world's first mining academy was established in the same town.

The widespread adoption of agricultural innovations such as the iron plowshare, as well as the growing use of metal as a building material, was also a driving force in the tremendous growth of the iron industry during this period. Inventions like the arrastra were often used by the Spanish to pulverize ore after being mined. This device was powered by animals and used the same principles used for grain threshing.

Much of the knowledge of medieval mining techniques comes from books such as Biringuccio's *De la pirotechnia* and probably most importantly from Georg Agricola's *De re metallica* (1556). These books detail many different mining methods used in German and Saxon mines. One of the prime issues confronting mediaeval miners (and one which Agricola explains in detail) was the removal of water from mining shafts. As miners dug deeper to access new veins, flooding became a very real obstacle. The mining industry became dramatically more efficient and prosperous with the invention of mechanical and animal driven pumps.

The Americas

There are ancient, prehistoric copper mines along Lake Superior, and metallic copper was still found there, near the surface, in colonial times. Indians availed themselves of this copper starting at least 5000 years ago," and copper tools, arrowheads, and other artifacts that were part of an extensive native trade network have been discovered. In addition, obsidian,flint, and other minerals were mined, worked, and traded. While the early French explorers that encountered the sites made no use of the metals due to the difficulties in transporting it, the copper was eventually traded throughout the continent along major river routes. In Saskatchewan, Canada, there also are ancient quartz mines near Waddy Lake and surrounding regions. In the early colonial history of the Americas, "native gold and silver was quickly expropriated and sent back to Spain in fleets of gold- and silver-laden galleons"mostly from mines in Central and South America. Turquoise dated at 700 A.D. was mined in pre-Columbian America; in the Cerillos Mining District in New Mexico, estimates are that "about 15,000 tons of rock had been removed from Mt Chalchihuitl using stone tools before 1700."

Mining in the United States became prevalent in the 19th century, and the General Mining Act of 1872 was passed to encourage mining of federal lands. As with the California Gold Rush in the mid 19th century, mining for minerals and precious metals, along with ranching, was a driving factor in the Westward Expansion to the Pacific coast. With the exploration of the West, mining camps were established and "expressed a distinctive spirit, an enduring legacy to the new nation;" Gold Rushers would experience the same problems as the Land Rushers of the transient West that preceded them.

Aided by railroads, many traveled West for work opportunities in mining. Western cities such as Denver and Sacramento originated as mining towns. As new areas were explored, it was usually the gold (placer and then load) and then silver that were taken first, with other metals often waiting for railroads or canals. Coarse gold dust and nuggets do not require smelting, is easy to identify and is easily transported.

Modern period

In the early 20th century, the gold and silver rush to the western United States also stimulated mining for base metals such as copper, lead, and iron as well as coal. Areas in modern Montana, Utah, Arizona, and later Alaska became predominate suppliers of copper to the world, which was increasingly demanding copper for electrical and households goods. Canada's mining industry grew more slowly than the United States due to limitations in transportation, capital, and U.S. competition; Ontario was the major producer of the early 20th century with nickel, copper, and gold.

Meanwhile, Australia experienced the Australian gold rushes and by the 1850s was producing 40% of the world's gold, followed by the establishment of large mines such as the Mount Morgan Mine, which ran for nearly a hundred years, Broken Hill ore deposit (one of the largest zinc-lead ore deposits), and iron ore mines at Iron Knob. After declines in production, another boom in mining occurred in the 1960s and in the 21st century Australia remains a major world mineral producer. Into the 21st century, a globalized mining industry of large multinational corporations has arisen. Peak minerals and environmental impacts have also become a concern. Different elements, particularly rare earth minerals, have begun to increase in demand as a result of new technologies.

Mine development and lifecycle

The process of mining from discovery of an ore body through extraction of minerals and finally to returning the land to its natural state consists of several distinct steps. The first is discovery of the ore body, which is carried out through prospecting or exploration to find and then define the extent, location and value of the ore body. This leads to a mathematical resource estimation to estimate the size and grade of the deposit. This estimation is used to conduct a pre-feasibility study to determine the theoretical economics of the ore deposit. This identifies, early on, whether further investment in estimation and engineering studies is warranted and identifies key risks and areas for further work. The next step is to conduct a feasibility study to evaluate the financial viability, technical and financial risks and robustness of the project.

This is when the mining company makes the decision to develop the mine or to walk away from the project. This includes mine planning to evaluate the economically recoverable portion of the deposit, the metallurgy and ore recoverability, marketability and payability of the ore concentrates, engineering concerns, milling and infrastructure costs, finance and equity requirements and an analysis of the proposed mine from the initial excavation all the way through to reclamation. The proportion of a deposit that is economically recoverable is dependent on the enrichment factor of the ore in the area.

To gain access to the mineral deposit within an area it is often necessary to mine through or remove waste material which is not of immediate interest to the miner. The total movement of ore and waste constitute the mining process. Often more waste than ore is mined during the life of a mine, depending on the nature and location of the orebody. Waste removal and placement is a major cost to the mining operator, so detailed characterization of the waste material forms an essential part of the geological exploration program for a mining operation.

Once the analysis determines a given ore body is worth recovering, development begins to create access to the ore body. The mine buildings and processing plants are built and any necessary equipment is obtained. The operation of the mine to recover the ore begins and continues as long as the company operating the mine finds it economical to do so. Once all the ore that the mine can produce profitably is recovered, reclamation begins to make the land used by the mine suitable for future use.

Mining techniques

Mining techniques can be divided into two common excavation types: surface mining and sub-surface (underground) mining. Today, surface mining is much more common, and produces, for example, 85% of minerals (excluding petroleum and natural gas) in the United States, including 98% of metallic ore.

Targets are divided into two general categories of materials: *placer deposits*, consisting of valuable minerals contained within river gravels, beach sands, and other unconsolidated materials; and *lode deposits*, where valuable minerals are found in veins, in layers, or in mineral grains generally distributed throughout a mass of actual rock. Both types of ore deposit, placer or lode, are mined by both surface and underground methods.

Fig. Underground longwall mining.

Some mining, including much of the rare earth elements and uranium mining, is done by less-common methods, such as in-situ leaching: this technique involves digging neither at the surface nor underground. The extraction of target minerals by this technique requires that they be soluble, *e.g.*, potash, potassium chloride, sodium chloride, sodium sulfate, which dissolve in water. Some minerals, such as copper minerals and uranium oxide, require acid or carbonate solutions to dissolve.

Surface mining

Surface mining is done by removing (stripping) surface vegetation, dirt, and if necessary, layers of bedrock in order to reach buried ore deposits. Techniques of surface mining include; Open-pit mining which consists of recovery of materials from an open pit in the ground, quarrying or gathering building materials from an open pit mine, strip mining which consists of stripping surface layers off to reveal ore/seams underneath, and mountaintop removal, commonly associated with coal mining, which involves taking the top of a mountain off to reach ore deposits at depth.

Most (but not all) placer deposits, because of their shallowly buried nature, are mined by surface methods. Landfill mining, finally, involves sites where landfills are excavated and processed.

Fig. Garzweiler surface mine, Germany

Underground mining

Fig. Mantrip used for transporting miners within an underground mine

Sub-surface mining consists of digging tunnels or shafts into the earth to reach buried ore deposits. Ore, for processing, and waste rock, for disposal, are brought to the surface through the tunnels and shafts. Sub-surface mining can be classified by the type of access shafts used, the extraction method or the technique used to reach the mineral deposit. Drift mining utilizes horizontal access tunnels, slope mining uses diagonally sloping access shafts and shaft mining consists of vertical access shafts. Mining in hard and soft rock formations require different techniques.

Other methods include shrinkage stope mining which is mining upward creating a sloping underground room, long wall miningwhich is grinding a long ore surface underground and room and pillar which is removing ore from rooms while leaving pillars in place to support the roof of the room. Room and pillar mining often leads to retreat mining which is removing the pillars which support rooms, allowing the room to cave in, loosening more ore. Additional sub-surface mining methods include hard rock miningwhich is mining of hard materials, bore hole mining, drift and fill mining, long hole slope mining, sub level caving and block caving.

Machinery

Fig. The Bagger 288 is a bucket-wheel excavator used instrip mining. It is also the largest land vehicle of all time.

Heavy machinery is used in mining for exploration and development, to remove and stockpile overburden, to break and remove rocks of various hardness and toughness, to process the ore and for reclamation efforts after the mine is closed. Bulldozers, drills, explosives and trucks are all necessary for excavating the land. In the case of placer mining, unconsolidated gravel, or alluvium, is fed into machinery consisting of a hopper and a shaking screen ortrommel which frees the desired minerals from the waste gravel. The minerals are then concentrated using sluices or jigs.

Large drills are used to sink shafts, excavate stopes and obtain samples for analysis. Trams are used to transport miners, minerals and waste. Lifts carry miners into and out of mines, as well as moving rock and ore out, and machinery in and out of underground mines. Huge trucks, shovels and cranes are employed in surface mining to move large quantities of overburden and ore. Processing plants can utilize large crushers, mills, reactors, roasters and other equipment to consolidate the mineral-rich material and extract the desired compounds and metals from the ore.

Processing

Once the mineral is extracted, it is often then processed. The science of extractive metallurgy is a specialized area in the science of metallurgy that studies the extraction of valuable metals from their ores, especially through chemical or mechanical means.

Mineral processing (or mineral dressing) is a specialized area in the science of metallurgy that studies the mechanical means of crushing, grinding, and washing that enable the separation (extractive metallurgy) of valuable metals or minerals from their gangue (waste material). Processing of placer ore material consists of gravity-dependent methods of separation, such as sluice boxes. Only minor shaking or washing may be necessary to disaggregate (unclump) the sands or gravels before processing. Processing of ore from a lode mine, whether it is a surface or subsurface mine, requires that the rock ore be crushed and pulverized before extraction of the valuable minerals begins. After lode ore is crushed, recovery of the valuable minerals is done by one, or a combination of several, mechanical and chemical techniques. Since most metals are present in ores as oxides or sulfides, the metal needs to be reduced to its metallic form. This can be accomplished through chemical means such assmelting or through electrolytic reduction, as in the case of aluminium. Geometallurgy combines the geologic sciences with extractive metallurgy and mining.

ENVIRONMENTAL EFFECTS

Environmental issues can include erosion, formation of sinkholes, loss of biodiversity, and contamination of soil, groundwater and surface water by chemicals from mining processes. In some cases, additional forest logging is done in the vicinity of mines to increase the available room for the storage of the created debris and soil. Contamination resulting from leakage of chemicals can also affect the health of the local population if not properly controlled. Extreme examples of pollution from mining activities include coal fires, which can last for years or even decades, producing massive amounts of environmental damage.

Fig. Iron hydroxide precipitate stains a stream receiving acid drainage from surface coal mining.

Mining companies in most countries are required to follow stringent environmental and rehabilitation codes in order to minimize environmental impact and avoid impacts on human health. These codes and regulations all require the common steps of environmental impact assessment, development of environmental management plans, mine closure planning (which must be done before the start of mining operations), and environmental monitoring during operation and after closure. However, in some areas, particularly in the developing world, regulation may not be well enforced by governments.

For major mining companies, and any company seeking international financing, there are however a number of other mechanisms to enforce good environmental standards. These generally relate to financing standards such as Equator Principles, IFC environmental standards, and criteria for Socially responsible investing. Mining companies have used this financial industry oversight to argue for some level of self-policing. In 1992 a Draft Code of Conduct for Transnational Corporations was proposed at the Rio Earth Summit by the UN Centre for Transnational Corporations (UNCTC), but the

Business Council for Sustainable Development (BCSD) together with the International Chamber of Commerce (ICC) argued successfully for self-regulation instead.

This was followed up by the Global Mining Initiative which was initiated by nine of the largest metals and mining companies, and led to the formation of the International Council on Mining and Metals to "act as a catalyst" for social and environmental performance improvement in the mining and metals industry internationally. The mining industry has provided funding to various conservation groups, some of which have been working with conservation agendas that are at odds with emerging acceptance of the rights of indigenous people - particularly rights to make land-use decisions.

Certification of mines with good practices occurs through the International Organization for Standardization (ISO) such as ISO 9000 and ISO 14001, which certifies an 'auditable environmental management system'; this certification involves short inspections, although it has been accused of lacking rigor. Certification is also available through Ceres' Global Reporting Initiative, but these reports are voluntary and unverified. Miscellaneous other certification programs exist for various projects, typically through nonprofit groups.

The purpose of a 2012 EPS PEAKS paper was to provide evidence on policies managing ecological costs and maximise socio-economic benefits of mining using host country regulatory initiatives. It found existing literature suggesting donors encourage developing countries to: Make the environment-poverty link and introduce cutting-edge wealth measures and natural capital accounts. Reform old taxes in line with more recent financial innovation, engage directly with the companies, enacting land use and impact assessments, and incorporate specialised support and standards agencies. Set in play transparency and community participation initiatives using the wealth accrued.

Waste

Ore mills generate large amounts of waste, called tailings. For example, 99 tons of waste are generated per ton of copper, with even higher ratios in gold mining. These tailings can be toxic. Tailings, which are usually produced as a slurry, are most commonly dumped into ponds made from naturally existing valleys. These ponds are secured by impoundments (dams or embankment dams). In 2000 it was estimated that 3,500 tailings impoundments existed, and that every year, 2 to 5 major failures and 35 minor failures occurred; for example, in the Marcopper mining disaster at least 2 million tons of tailings were released into a local river. Subaqueous tailings disposal is another option. The mining industry has argued that submarine tailings disposal (STD), which disposes of tailings in the sea, is ideal because it avoids the risks of tailings ponds; although the practice is illegal in the United States and Canada, it is used in the developing world. The waste is classified as either sterile or mineralised, with acid generating potential, and the movement and storage of this material forms a major part of the mine planning process. When the mineralised package is determined by an economic cut-off, the near-grade mineralised waste is usually dumped separately with view to later treatment should market conditions change and it becomes economically viable. Civil engineering design parameters are used in the design of the waste dumps, and special conditions apply to high-rainfall areas and to seismically active areas. Waste dump designs must meet all regulatory requirements of the country in whose jurisdiction the mine is located. It is also common practice to rehabilitate dumps to an internationally acceptable standard, which in some cases means that higher standards than the local regulatory standard are applied.

MINING INDUSTRY

Mining exists in many countries. London is known as the capital of global "mining houses" such as Rio Tinto Group, BHP Billiton, and Anglo American PLC. The US mining industry is also large but it is dominated by the coal and other nonmetal minerals (*e.g.*, rock and sand), and the various regulations have worked to reduce the significance of mining in the United States. In 2007 the total market cap of mining companies was reported at US$962 billion, which compares to a total global market cap of publicly traded companies of about US$50 trillion in 2007. In 2002, Chile and Peru were reportedly the major mining countries of South America. The mineral industry of Africaincludes the mining of various minerals; it produces relatively little of the industrial metals copper, lead and zinc but according to one estimate has as a percent of world reserves 40% of gold, 60% cobalt and 90% of the world's platinum group metals.

Mining in India is a significant part of that country's economy. In the developed world,mining in Australia with BHP Billiton founded and headquartered in the country, and Canada are particularly significant. For rare earth minerals mining, China reportedly controlled 95% of production in 2013.While exploration and mining can sometimes be conducted by individual entrepreneurs or small business, most modern-day mines are large enterprises requiring large amounts of capital to establish. Consequently, the mining sector of the industry is dominated by large, often multinational companies, most of them publicly listed.

It can be argued that what is referred to as the 'mining industry' is actually two sectors, one specializing in exploration for new resources, the other specializing in mining those resources. The exploration sector is typically made up of individuals and small mineral resource companies ("juniors") dependent on venture capital. The mining sector is typically large and multi-national companies sustained by mineral production from their mining operations. In addition to these two sectors, various other industries such as equipment manufacture, environmental testing and metallurgy analysis also rely on and support the mining industry throughout the world.

Canadian stock exchanges have a particular focus on mining companies, particularly junior exploration companies through Toronto'sTSX Venture Exchange; Canadian companies raise capital on these exchanges and then invest the money in exploration globally. Some have argued that below juniors there exists a substantial sector of illegitimate companies primarily focused on manipulating stock prices.

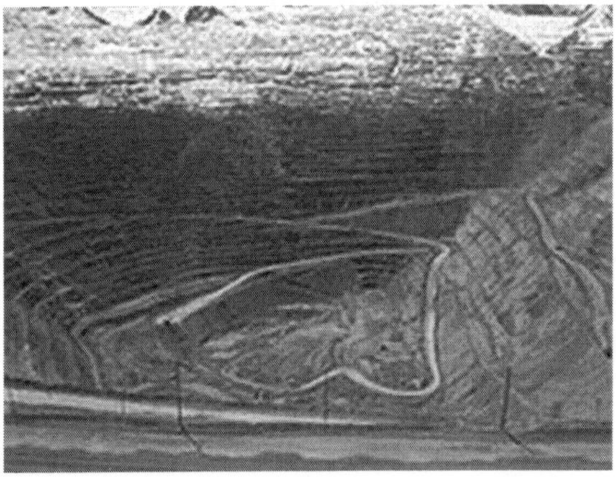

Fig. The Bingham Canyon Mine of Rio Tinto's subsidiary, Kennecott Utah Copper.

Mining operations can be grouped into five major categories in terms of their respective resources. These are, oil and gas extraction, coal mining, metal ore mining, nonmetallic mineral mining and quarrying, and support activities for mining. Out of all these categories, oil and gas extraction remains one of the largest in terms of its global economic importance. Prospecting potential mining sites, a vital area of concern for the mining industry is now done using sophisticated new technologies such as seismic prospecting and remote-sensing satellites.

Mining is heavily affected by the price of the commodity minerals, which is often volatile. The 2000s commodities boom ("commodities supercycle") increased the prices of commodities, driving aggressive mining. In addition, the price of gold increased dramatically in the 2000s, which increased gold mining; for example, one study found that conversion of forest in the Amazon increased six-fold from 2003–2006 (292 ha/yr) to 2006–2009 (1915 ha/year) largely due to artisanal mining.

Corporate classifications

Mining companies can be classified based on their size and financial capabilities:

- Major companies are considered to have an adjusted annual mining-

related revenue of more than US$500 million, with the financial capability to develop a major mine on its own.
- Intermediate companies have at least $50 million in annual revenue but less than $500 million.
- Junior companies rely on equity financing as their principal means of funding exploration. Juniors are mainly pure exploration companies, but may also produce minimally, and do not have a revenue exceeding US$50 million.

Regulation and governance

New regulation and process of legislative reforms aims to enrich the harmonization and stability of the mining sector in mineral-rich countries. The new legislation for mining industry in the African countries still appears as an emerging issue with a potential to be solved, until a consensus is reached on the best approach. By the beginning of 20th century the booming and more complex mining sector in mineral-rich countries provided only slight benefits to local communities in terms of sustainability. Increasing debates and influence by NGOs and communities appealed for a new program which would have had also included a disadvantaged communities, and would have had worked towards sustainable development even after mine closure (included transparency and revenue management). By the early 2000s, community development issues and resettlements became mainstreamed in Bank mining projects. Mining-industry expansion after an increase of mineral prices in 2003 and also potential fiscal revenues in those countries created an omission in the other economic sectors in terms of finances and development.

Furthermore, it had highlighted regional and local demand of mining-revenues and lack of ability of sub-national governments to use the revenues. The Fraser Institute (a Canadian think tank) has highlighted the environmental protection laws in developing countries, as well as the voluntary efforts by mining companies to improve their environmental impact. In 2007 the Extractive Industries Transparency Initiative (EITI) was mainstreamed in all countries cooperating with the World Bank in mining industry reform. The EITI is operating and implementing with a support of EITI Multi-Donor Trust Fund, managed by The World Bank. The Extractive Industries Transparency Initiative (EITI) aims to increase transparency in transactions between governments and companies within extractive industries by monitoring the revenues and benefits between industries and recipient governments.

The entrance process is voluntary for each country and is being monitored by multi-stakeholders involving government, private companies and civil society representatives, responsible for disclosure and dissemination of the reconciliation report; however, the competitive disadvantage of company-by company public report is for some of the businesses in Ghana, the main constraint. Therefore, the outcome assessment in terms of failure or success of the new EITI regulation does not only "rest on the government's shoulders" but also on civil society and companies. On the other hand, criticism points out two main implementation issues; inclusion or exclusion of artisanal mining and small-scale mining (ASM) from the EITI and how to deal with "non-cash" payments made by companies to subnational governments. Furthermore, disproportion of the revenues mining industry creates to the comparatively small number of people that it employs, causes another controversy.

The issue of artisanal mining is clearly an issue in EITI Countries such as the Central African Republic, D.R. Congo, Guinea, Liberia and Sierra Leone – *i.e.* almost half of the mining countries implementing the EITI. Among other things, limited scope of the EITI involving disparity in terms of knowledge of the industry and negotiation skills, thus far flexibility of the policy (*e.g.* liberty of the countries to expand beyond the minimum requirements and adapt it to their needs), creates another risk of unsuccessful implementation. Public awareness increase, where government should act as a bridge between public and initiative for a successful outcome of the policy is an important element to be considered.

World Bank

The World Bank has been involved in mining since 1955, mainly through grants from its International Bank for Reconstruction and Development, with the Bank's Multilateral Investment Guarantee Agency offering political risk insurance.

Between 1955 and 1990 it provided about $2 billion to fifty mining projects, broadly categorized as reform and rehabilitation, greenfield mine construction, mineral processing, technical assistance, and engineering. These projects have been criticized, particularly the Ferro Carajas project of Brazil, begun in 1981. The bank established mining codes intended to increase foreign investment, in 1988 solicited feedback from 45 mining companies on how to increase their involvement.

In 1992 the bank began to push for privatization of government-owned mining companies with a new set of codes, beginning with its report *The Strategy for African Mining*. In 1997, Latin America's largest miner Companhia Vale do Rio Doce (CVRD) was privatized. These and other movements such as the Philippines 1995 Mining Act led the World Bank to publish a third report (*Assistance for Minerals Sector Development and Reform in Member Countries*) which endorsed mandatory environment impact assessments and attention to the locals.

The codes based on this report are influential in the legislation of developing nations. The new codes are intended to encourage development through tax holidays, zero custom duties, reduced income taxes, and related measures. The results of these codes were analyzed by a group from the University of Quebec, which concluded that the codes promote foreign investment but "fall very short of permitting sustainable development". The observed negative correlation between natural resources and economic development is known as the resource curse.

SAFETY

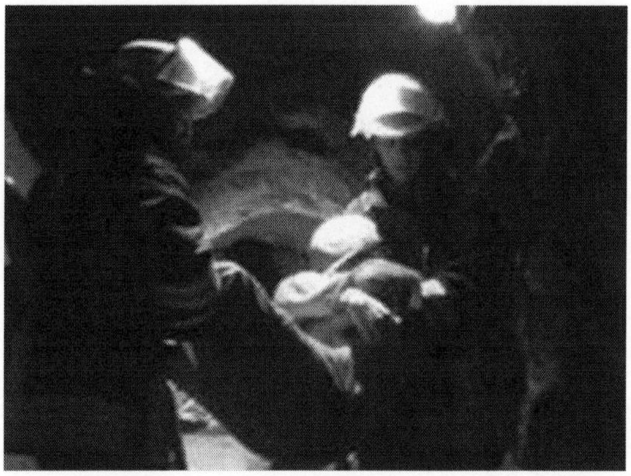

Fig. Firefighter training in fell slate mine,Germany

Fig. Remnants of pit "San Vicente" inLinares, Jaén, Spain. On its last day of operation, in 1967, six workers died in anelevator accident.

Safety has long been a concern in the mining business especially in sub-surface mining. The Courrières mine disaster, Europe's worst mining accident, involved the death of 1,099 miners in Northern France on 10 March 1906. This disaster was surpassed only by the Benxihu Colliery accident in China on April 26, 1942, which killed 1,549 miners. While mining today is substantially safer than it was in the previous decades, mining accidents still occur. Government figures indicate that 5,000 Chinese miners die in accidents each year, while other reports have suggested a figure as high as 20,000. Mining accidents continue worldwide, including accidents causing dozens of fatalities at a time, such as the 2007 Ulyanovskaya Mine disaster in Russia, the 2010

Upper Big Branch Mine disaster in the United States, and the 2009 Heilongjiang mine explosion in China. Mining ventilation is a significant safety concern for many miners. Poor ventilation of the mines causes exposure to harmful gases, heat and dust inside sub-surface mines, which can cause injury, illness and death. The concentration of methane and other airborne contaminants underground can generally be controlled by dilution (ventilation), capture before entering the host air stream (methane drainage), or isolation (seals and stoppings).

Rock dusts, including coal dust and silicon dust can cause long-term lung problems, including silicosis, asbestosis and pneumoconiosis (also known as miners lung or black lung disease). A ventilation system is set up to force a stream of air through the working areas of the mine. The air circulation necessary for the effective ventilation of a mine is generated by one or more large mine fans, usually located above ground. Air flows in one direction only, making circuits through the mine such that each main work area constantly receives a supply of fresh air.

Watering down in coal mines also helps to keep dust levels down: by spraying the machine with water and filtering the dust-laden water with a scrubber fan, miners can successfully trap the dust. Gases in mines can poison the workers or displace the oxygen in the mine, causing asphyxiation. For this reason, the U.S. Mine Safety and Health Administration requires that groups of miners in the United States carry gas detection equipment that can detect common gases, such as CO, O_2, H_2S, CH_4, as well as calculate % Lower Explosive Limit. Regulation requires that all production stop if there is a concentrate of 1.4% of flammable gas present. Additionally, further regulation is being requested for more gas detection as newer technology such as nanotechnology is introduced.

Ignited methane gas is a common source of explosions in coal mines, which can potentially initiate more extensive coal dustexplosions. For this reason, rock dusts such as limestone dust are spread throughout coal mines to diminish the chances of coal dust explosions as well as to limit the extent of potential explosions, in a process known as rock dusting. Coal dust explosions can also begin independently of methane gas explosions. Frictional heat and sparks generated by mining equipment can ignite both methane gas and coal dust explosions. For this reason, water is often used to cool rock cutting sites.

Miners utilize equipment strong enough to break through extremely hard layers of the Earth's crust. This equipment, combined with the closed workspace that underground miners work in, can cause hearing loss. For example, a roof bolter (commonly used by mine roof bolter operators) can reach sound power levels of up to 115 dB. Combined with the reverberant effects of underground mines, a miner without proper hearing protection is at a high risk for hearing loss. By age 50, nearly 90% of U.S. coal miners have some hearing loss, compared to only 10% among workers not exposed to loud noises. Roof bolters are among the loudest machines, but auger miners, bulldozers, continuous mining machines, front end loaders, and shuttle cars and trucks are also among those machines most responsible for excessive noise in mine work.

Since mining entails removing dirt and rock from its natural location creating large empty pits, rooms and tunnels, cave-ins as well as ground and rock falls are a major concern within mines. Modern techniques for timbering and bracing walls and ceilings within sub-surface mines have reduced the number of fatalities due to cave-ins, but ground falls continue to represent up to 50% of mining fatalities.

Even in cases where mine collapses are not instantly fatal, they can trap mine workers deep underground. Cases such as these often provoke high-profile rescue efforts, such as when 33 Chilean miners were trapped deep underground for 69 days in 2010.

High temperatures and humidity may result in heat-related illnesses, including heat stroke which can be fatal. The presence of heavy equipment in confined spaces also poses a risk to miners, and despite modern improvements to safety practices, mining remains dangerous throughout the world. Modern mines use automation and remote operation to improve the safety of mine workers. Examples include automated loaders and remotely operated rockbreakers.

Abandoned mines

Fig. Warning sign near a dangerous area filled with open mineshafts, Calico Ghost Town, California.

There are upwards of 560,000 abandoned mines on public and privately owned lands in the United States alone.Abandoned mines may be dangerous to anyone who attempts to explore them without proper knowledge and safety training.

RECORDS

Fig. Chuquicamata, Chile, site of the largest circumference and second deepest open pit copper mine in the world.

As of 2008, the deepest mine in the world is TauTona in Carletonville, South Africa at 3.9 kilometers, replacing the neighboringSavuka Mine in the North West Province of South Africa at 3,774 meters. East Rand Mine in Boksburg, South Africa briefly held the record at 3,585 meters, and the first mine declared the deepest in the world was also TauTona when it was at 3,581 meters. The Moab Khutsong gold mine in North West Province (South Africa) has the world's longest winding steel wire rope, able to lower workers to 3 054 metres in one uninterrupted four-minute journey.

The deepest mine in Europe is the 16th shaft of the uranium mines in Pøíbram, Czech Republic at 1,838 meters, second isBergwerk Saar in Saarland, Germany at 1,750 meters. The deepest open pit mine in the world is Bingham Canyon Mine in Bingham Canyon, Utah, United States at over 1,200 meters. The largest and second deepest open pit copper mine in the world is Chuquicamata in Chuquicamata, Chile at 900 meters, 940,600 tons of copper and 17,700 tons of molybdenum produced annually.

The deepest open pit mine with respect to sea level is Tagebau Hambach in Germany, where the base of the pit is 293 meters below sea level. The largest underground mine is Kiirunavaara Mine in Kiruna, Sweden. With 450 kilometers of roads, 40 million tonnes of ore produced yearly, and a depth of 1,270 meters, it is also one of the most modern underground mines. The deepest borehole in the world is Kola Superdeep Borehole at 12,262 meters. This, however, is not a matter of mining but rather related to scientific drilling.

METAL RESERVES AND RECYCLING

During the 20th century, the variety of metals used in society grew rapidly. Today, the development of major nations, such as China and India, and advances in technologies, are fuelling ever more demand. The result is that metal mining activities are expanding, and more and more of the world's metal stocks are above ground in use, rather than below ground as unused reserves. An example is the in-use stock of copper. Between 1932 and 1999, copper in use in the USA rose from 73 kg to 238 kg per person.

95% of the energy used to make aluminium from bauxite ore is saved by using recycled material. However, levels of metals recycling are generally low. In 2010, theInternational Resource Panel, hosted by the United Nations Environment Programme (UNEP) published reports on metal stocks that exist within society and their recycling rates.

The report's authors observed that the metal stocks in society can serve as huge mines above ground. However, they warned that the recycling rates of some rare metals used in applications such as mobile phones, battery packs for hybrid cars and fuel cells, are so low that unless future end-of-life recycling rates are dramatically stepped up these critical metals will become unavailable for use in modern technology.

LANDFILL MINING

Landfill mining and reclamation (LFMR) is a process whereby solid wastes which have previously been landfilled are excavated and processed. The function of landfill mining is to reduce the amount of landfill mass encapsulated within the closed landfill and/or temporarily remove hazardous material to allow protective measures to be taken before the landfill mass is replaced. In the process, mining recovers valuable recyclable materials, a combustible fraction, soil, and landfill space. The aeration of the landfill soil is a secondary benefit regarding the landfill's future use. The combustible fraction is useful for the generation of power. The overall appearance of the landfill mining procedure is a sequence of processing machines laid out in a functional conveyor system. The operating principle is to excavate, sieve and sort the landfill material.

The concept of landfill mining was introduced as early as 1953 at the Hiriya landfill operated by the Dan Region Authority next to the city of Tel Aviv, Israel. Waste contains many resources with high value, the most notable of which are non-ferrous metals such as aluminium cans and scrap metal. The concentration of aluminium in many landfills is higher than the concentration of aluminum in bauxite from which the metal is derived.

Practical applications

Landfill mining is also possible in countries where land is not available for new landfill sites. In this instance landfill space can be reclaimed by the extraction of biodegradable waste and other substances then refilled with wastes requiring disposal. Mining construction landfill sites is the simplest form of landfill mining. Construction landfills contain three basic components, wood, scrap metal and gypsum, or drywall, along with a minimal amount of other construction materials. The wood collected can be used as fuel in coal burning power plants and the scrap metal reprocessed.

Mining of municipal landfills is more complicated and has to be based on the expected content of the landfill. Older landfills, in the United States before 1994, were often capped and closed, essentially entombing the waste. This can be beneficial for waste recovery. It can also create a higher risk for toxic waste and leachate exposure as the landfill has not fully processed the stewing wastes. Mining of bioreactor landfills and properly stabilized modern sanitary landfills provides its own benefits. The biodegradable wastes are more easily sieved out, leaving the non-biodegradable materials readily accessible. The quality of these materials for recycling and reprocessing purposes is not as high as initially recycled materials, however materials such as aluminum and steel are usually excluded from this.

Landfill mining is most useful as a method to remediate hazardous landfills. Landfills that were established before landfill liner technology was well established often leak their unprocessed leachate into underlying aquifers. This is both an environmental hazard and also a legal liability. In the US, Environmental Protection Agency fines can tax the local economy up to 30 years after the site has closed. Mining the landfill simply to lay a safe liner is a last, but sometimes necessary resort.

Tools and machinery

The parts of the mining process are the different mining machines. Depending on the complexity of the process more or fewer machines can be used. Machinery is easily transported on trucks from site to site, mounted on trailers. The following machines are added in order in increase of mining complexity:

- Excavators
- Moving floor and elevator conveyor belts
- A coarse rotating trommel screen
- A fine rotating trommel screen
- A magnet
- Front end loader
- Odor control sprayer

The mechanics of mining

An excavator or front end loader uncovers the landfilled materials and places them on a moving floor conveyor belt to be taken to the sorting machinery. A trommel is used to separate materials by size. First, a large trommel separates materials like appliances and fabrics. A smaller trommel then allows the biodegraded soil fraction to pass through leaving non-biodegradable, recyclable materials on the screen to be collected. An electromagnet is used to remove the ferrous material from the waste mass as it passes along the conveyor belt. A front end loader is used to move sorted materials to trucks for further processing. Odour control sprayers are wheeled tractors with a cab and movable spray arm mounted on a rotating platform. A large reservoir tank mounted behind the cab holds neutralising agents, usually in liquid form, to reduce the smell of exposed wastes.

Operational flow

Excavators dig up waste mass and transport it, with the help of front end loaders, onto elevator and moving floor conveyor belts. The conveyor belts empty into a coarse, rotating trommel. The large holes in the screen allow most wastes to pass through, leaving behind the oversized, non-processable materials. The oversized wastes are removed from inside the screen. The coarse trommel empties into the fine rotating trommel. The fine rotating trommel allows the soil fraction to pass through, leaving mid-sized, non-biodegradable, mostly recyclable materials. The materials are removed from the screen. These materials are put on a second conveyor belt where an electromagnet removes any ferromagnetic debris. Depending on the level of resource recovery, material can be put through an air classifier which separates light organic material from heavy organic material. The separate streams are then loaded, by front end loaders, onto trucks either for further processing or for sale. Further manual processing can be done on site if processing facilities are too far away to justify the transportation costs.

MINE DEVELOPMENT AND PRODUCTION

Mine production will proceed with the advance of trenches, each trench at different floor level maintaining close to floor. Each trench will have about 80- to 100-m width. The overburden front is mostly planned to have an advance of about 80-100 m ahead of bauxite front, overburden excavation is carried out in single bench of about 8-m on an average, however bauxite excavation is proposed to be carried out in two slices, the top slice (about 10-m thick) will involve unhindered mass-mining, the bottom slice, usually about 4 to 5-m thick, is required to be mined selectively to exclude the non-bauxite high silica zones on the floor. At places with high bauxite thickness, the unhindered mining of bauxite layer may be mined in two slices, so as to leave only about 4 to 5-m or even less for selective mining.

The area around first mine cut will be developed to a width of about 1000-m, so as to have about 10-12 trenches advancing parallel to one another. The semi-mobile crusher will be set up in the central trench so as to have equal distance from the farthest trench. The second crusher will be located during the seventh year of mine operation on the Northern part of the deposit. From 9 year onwards the two crushers together will deliver full production of 8.5 MTPA of bauxite to the alumina plant. Each crusher will have a capacity of 4.25 MTPA. The mine development involving excavation of overburden and bauxite for initial period of five years. The overburden excavated during the first three years is totally dumped on the southern plateau top. The overburden excavated in the fourth year is partly (2.10-MT) dumped in the designated OB dumps and subsequently backfilled into mined out areas. The year-wise detailed development and production schemes are described in the following sections.

Table. Yearly Excavation of Overburden and Bauxite

Year of operation	Overburden Removal (MT)	Bauxite Mined (MT)	Grade of Bauxite (%)		Cumulative Bauxite (MT)
			Al_2O_3	SiO_2	
1	2.14	1.24	44.21	2.02	1.24
2	1.00	1.60	44.63	2.10	2.84
3	2.15	3.08	44.34	2.13	5.92
4	3.10	3.70	44.43	2.14	9.62
5	3.02	3.98	44.78	2.12	13.60
6-8	21.78	23.94	44.46	2.16	37.54
9	6.95	8.50	44.28	2.29	46.04

Year-wise Development and Production for the First Five Years of Operation.

Year-wise detailed mine development drawings and excavation schedules are prepared for the first 5 years. During the first five years 67-ha of area is excavated of which 15-ha area will be backfilled and reclaimed. In the first five years total of 11.41 MT of overburden will be excavated and during the same period 13.60 MT of bauxite will also be excavated. Conceptual mine development drawings at the end of 5 year of operations.

Mine Development during First Year

During first year, apart from opening of the deposit, various construction activities of buildings such as office, workshop, other facilities area, access road, electrical transmission line and distribution network, water pumping, supply, storage and distribution etc will be completed. During this period, 2.14 MT of overburden will be excavated from an area of 14-ha and the overburden so excavated will be dumped in a designated area on the southern part of the plateau top. This will expose about 3.0 MT of bauxite for excavation from different mine faces. Incidental bauxite that will be excavated during the first year will be 1.24 MT, which will be stocked close to the proposed Semi Mobile Crusher. During this period a Semi-Mobile Crusher also will be installed at the floor of bottom bauxite such that the hopper level and the floor of top bauxite are at the same level.

Mine Development during Second Year

The second year of mine development coincides with the Alumina plant commissioning. Excavation is carried out around the initial mine opening during the second year. The quantum of bauxite excavation during second year will be 1.6-MT. This, together with the bauxite stockpiled during the first year (1.24-MT) totaling to 2.84-MT, will be crushed and conveyed to Alumina plant through a long distance conveyor.

The overburden excavated during second year will be 1.00-MT with mine fronts advancing westwards. The overburden excavated is planned to be dumped as in the first year on the southern part of the plateau top since, sufficient mined out area is still not available for backfilling of overburden. Additional 7.0-ha area is exposed during the second year totaling to 21-ha (including 14-ha of the first year).

Mine Development during Third Year

By this period, the Mine will be sufficiently developed to excavate the desired capacity of bauxite. Excavation and mine advancement progress towards west in parallel trenches following the bauxite floor. The quantum of bauxite excavation during the third year is 3.08-MT, which will be crushed and conveyed to Alumina plant through the long distance conveyor.

The overburden excavated during the third year will be 2.15-MT with mine fronts advancing westwards. The overburden excavated is planned to be dumped as in the earlier years on the southern part of the plateau top since, sufficient mined out area is still not available for backfilling of overburden. Additional 14-ha area is exposed during the third year totaling to 35-ha (including 21-ha of the first two years).

Mine Development during Fourth Year

Excavation and mine advancement continues in parallel trenches towards west following the bauxite floor. The quantum of bauxite excavation during the fourth year is 3.70-MT, which will be crushed and conveyed to Alumina plant through the long distance conveyor. The overburden excavated during the fourth year will be 3.10-MT with mine fronts advancing westwards. The overburden excavated is planned to be dumped as in the earlier years on the southern part of the plateau top. However, during the later part of fourth year sufficient bottom bauxite has been excavated and mined out area is available for backfilling.

Thus, the overburden excavated from then onwards will be backfilled into mined out areas. Additional 19-ha area is exposed during the fourth year totaling to 54-ha (including 35-ha of the first three years). Land reclamation of mined out area by backfilling is carried out over an area of 5-ha during the fourth year.

Mine Development during Fifth Year

Excavation and mine advancement continues in parallel trenches towards West following the bauxite floor. The quantum of bauxite excavation during the fifth year is 3.98-MT that will be crushed and conveyed to Alumina plant through the long distance conveyor. The overburden excavated during the fifth year will be 3.02-MT with mine fronts advancing westwards. The overburden excavated is planned to be dumped for backfilling into mined out areas available after the excavation of bottom bauxite. Additional 13-ha area is exposed during the fifth year totaling to 67-ha (including 54-ha of the first four years). Land reclamation of mined out area by backfilling is carried out over an area of 10-ha during the fifth year, thus cumulative area reclaimed at the end of fifth year will be 15-ha.

PROPOSED RATE OF PRODUCTION, WHEN MINE IS FULLY-DEVELOPED

The proposed rate of production of the mine, when the mine is fully developed is 8.5-MTPA of bauxite. The bauxite mine production will increase from the initial 1.24-MTPA to 8.5-MTPA during the 9 year to suit the requirement of alumina plant. The Semi Mobile Crusher Plant–I (SMCP-I), which is designed for production of 4.25-MTPA and is commissioned during second year, will be in the central part. To cater to the increase of bauxite excavation beyond 4.25-MTPA, a second Semi Mobile Crusher Plant–II (SMCP-II) of similar type and capacity is planned and commissioned on the northern part of the deposit, during the seventh year. During the seventh year the mined out area is sufficient to re-handle the overburden from OB Dumps and backfill into the mined out area around SMCP–I, in addition the overburden excavated around the SMCP–II area on the northern part will also be backfilled into the area available around SMCP–I.

In subsequent years, the bauxite excavation from northern part will further increase and in the ninth year both the working areas near the two SMCPs will be producing 8.5-MTPA of bauxite, simultaneously excavating about 7.0 MTPA of overburden.

CONCEPTUAL PRODUCTION PLAN FOR TENTH YEAR ONWARDS

Mine development/excavation will be stabilized from 9 year onwards to excavate 8.5-MTPA of bauxite. Mine Closure plan at 27 year of working. The mined out area corresponding to Crusher-I will initially advance westwards and then towards south. Accordingly, the crusher is shifted to a location on the Southern part of the plateau. Crusher-I will thus be handling all bauxite in Central, Eastern and Southern side of plateau and Crusher II will be handling all bauxite on the Northern and Western part of the plateau. Due care has been taken during mine planning for proper blending from both mining areas to meet the plant feed grade of bauxite. Year wise Mine plans prepared indicate the area excavated in sequence for overburden excavation, mass mining (of top bauxite), selective mining (of bottom bauxite) and area backfilled by overburden waste.

The block krigged values provide data on estimates of overburden thickness, bauxite thickness, quantity, its grade etc for each block of 25-mx25-m. Almost equal quantity of bauxite needs to be excavated from both the mining areas, depending on quality of each 25-mx25-m blocks, areas are selected for mining each year in a sequence so that the required plant feed grade is met.

DISCIPLINE MINING ENGINEERING

Mining engineering is an engineering discipline that involves the practice, the theory, the science, the technology, and application of extracting and processing minerals from a naturally occurring environment. Mining engineering also includes processing minerals for additional value. Mineral extraction is essential to modern society. Mining activities by their nature cause a disturbance of the environment in and around which the minerals are located. Mining engineers must therefore be concerned not only with the production and processing of mineral commodities, but also with the mitigation of damage to the environment as a result of that production and processing.

HISTORY OF MINING ENGINEERING

Since the beginning of civilization people have used stone and ceramics and, later, metals found on or close to the Earth's surface. These were used to manufacture earlytools and weapons. For example, high quality flint found in northern France and southern England were used to set fire and break rock. Flint mines have been found inchalk areas where seams of the stone were followed underground by shafts and galleries. The oldest known mine on archaeological record is the "Lion Cave" in Swaziland. At this site, which radiocarbon dating indicates to be about 43,000 years old, paleolithic humans mined mineral hematite, which contained iron and was ground to produce the red pigment ochre.

The ancient Romans were innovators of mining engineering. They developed large scale mining methods, perhaps most notably the use of large volumes of water brought to the minehead by numerous aqueducts for hydraulic mining. The exposed rock was then attacked by fire-setting where fires were used to heat the rock, which would be quenched with a stream of water. The thermal shock cracked the rock, enabling it to be removed. In some mines the Romans utilized water-powered machinery such asreverse overshot water-wheels. These were used extensively in the copper mines at Rio Tinto in Spain, where one sequence comprised 16 such wheels arranged in pairs, lifting water about 80 feet (24 m). Black powder was first used in mining in Banská Štiavnica, Kingdom of Hungary (present-day Slovakia) in 1627. This allowed blasting of rock and earth to loosen and reveal ore veins, which was much faster than fire-setting. The Industrial Revolution saw further advances in mining technologies, including improved explosives and steam-poweredpumps, lifts, and drills as long as they remained safe.

Salary and statistics

In the United States, there are an estimated 6,630 employed mining engineers. The mean yearly salary for a mining engineer in the U.S. is $90,070.

MINERAL EXPLORATION

Mining engineers are consulted for virtually every stage of a mining operation. The first role of engineering in mines is the discovery of a mineral deposit and the determination of the profitability of a mine.

Mineral discovery

Mining engineers are involved in the mineral discovery stage by working with geologists to identify a mineral reserve. The first step in discovering an ore body is to determine what minerals to test for. Geologists and engineers drill core samples and conduct surface surveys searching for specific compounds and ores. For example, a mining engineer and geologist may target metallic ores such as galena for lead or chalcopyrite for copper. A mining engineer may also search for a non-metal such as phosphate, quartz, or coal. The discovery can be made from research of mineral maps, academic geological reports or local, state, and national geological reports. Other sources of information include property assays, well drilling logs, and local word of mouth. Mineral research may also include satellite and airborne photographs. Unless the mineral exploration is done on public property, the owners of the property may play a significant role in the exploration process, and may be the original discoverer of the mineral deposit.

Mineral determination

After a prospective mineral is located, the mining engineer then determines the ore properties. This may involve chemical analysis of the ore to determine the composition of the sample. Once the mineral properties are identified, the next step is determining the quantity of the ore. This involves determining the extent of the deposit as well as the purity of the ore. The engineer drills additional core samples to find the limits of the deposit or seam and calculates the quantity of valuable material present in the deposit.

Feasibility Study

Once the mineral identification and reserve amount is reasonably determined, the next step is to determine the feasibility of recovering the mineral deposit. A preliminary study shortly after the discovery of the deposit examines the market conditions such as the supply and demand of the mineral, the amount of ore needed to be moved to recover a certain quantity of that mineral as well as analysis of the cost associated with the operation.

This pre-feasibility study determines whether the mining project is likely to be profitable; if it is then a more in-depth analysis of the deposit is undertaken. After the full extent of the ore body is known and has been examined by engineers, the feasibility study examines the cost of initial capital investment, methods of extraction, the cost of operation, an estimated length of time to payback, the gross revenue and net profit margin, any possible resale price of the land, the total life of the reserve, the total value of the reserve, investment in future projects, and the property owner or owners' contract.

In addition, environmental impact, reclamation, possible legal ramifications and all government permitting are considered. These steps of analysis determine whether the mine company should proceed with the extraction of the minerals or whether the project should be abandoned. The mining company may decide to sell the rights to the reserve to a third party rather than develop it themselves, or the decision to proceed with extraction may be postponed indefinitely until market conditions become favorable.

MINING OPERATION

Mining engineers working in an established mine may work as an engineer for operations improvement, further mineral exploration, and operation capitalization by determining where in the mine to add equipment and personnel. The engineer may also work in supervision and management, or as an equipment and mineral salesperson. In addition to engineering and operations, the mining engineer may work as an environmental, health and safety manager or design engineer.

The act of mining required different methods of extraction depending on the mineralogy, geology, and location of the resources. Characteristics such as mineral hardness, the mineral stratification, and access to that mineral will determine the method of extraction. Generally, mining is either done from the surface or underground. Mining can also occur with both surface and underground operations taking place on the same reserve. Mining activity varies as to what method is employed to remove the mineral.

Surface mining

Surface comprises 90% of the world's mineral tonnage output. Also called open pit mining, surface mining is removing minerals in formations that are at or near the surface. Ore retrieval is done by material removal from the land in its natural state. Surface mining often alters the land characteristics, shape, topography, and geological make-up.

Surface mining involves quarrying which is excavating minerals by means of machinery such as cutting, cleaving, and breaking. Explosives are usually used to facilitate breakage. Hard minerals such as limestone, sand, gravel, and slate are generally quarried into a series of benches. Strip mining is done on softer minerals such as clays and phosphate are removed through use of mechanical shovels, track dozers, and front end loaders. Softer Coal seams can also be extracted this way.

With placer mining, minerals can also be removed from the bottoms of lakes, rivers, streams, and even the ocean by dredge mining. In addition, in-situ mining can be done from the surface using dissolving agents on the ore body and retrieving the ore via pumping. The pumped material is then set to leach for further processing. Hydraulic mining is utilized in forms of water jets to wash away either overburden or the ore itself.

Mining process

Blasting

Explosives are used to break up a rock formation and aid in the collection of ore in a process called blasting. Blasting utilizes the heat and immense pressure of the detonated explosives to shatter and fracture a rock mass. The type of explosives used in mining are high explosives which vary in composition and performance properties. The mining engineer is responsible for the selection and proper placement of these explosives, in order to maximize efficiency and safety. Blasting occurs in many phases of the mining process, such as development of infrastructure as well as production of the ore.

MINING HEALTH AND SAFETY

Legal attention to Mining Health and Safety began in the late 19th century and in the subsequent 20th century progressed to a comprehensive and stringent codification of enforcement and mandatory health and safety regulation. A mining engineer in whatever role they occupy must follow all federal, state, and local mine safety laws.

United States

The United States Congress, through the passage of the Federal Mine Safety and Health Act of 1977, known as the Miner's Act, created the Mine Safety and Health Administration (MSHA) under the US Department of Labor. This comprehensive Act provides miners with rights against retaliation for reporting violations, consolidated regulation of coal mines with metallic and nonmetallic mines, and created the independent Federal Mine Safety and Health Review Commission to review MSHA's reported violations. The Act as codified in Code of Federal Regulations § 30 (CFR § 30) covers all miners at an active mine. When a mining engineer works at an active mine he or she is subject to the same rights, violations, mandatory health and safety regulations, and mandatory training as any other worker at the mine. The mining engineer can be legally identified as a "miner." The Act establishes the rights of miners. The miner may report at any time a hazardous condition and request an inspection. The miners may elect a miners' representative to participate during an inspection, pre-inspection meeting, and post-inspection conference. The miners and miners' representative shall be paid for their time during all inspections and investigations.

MINING AND THE ENVIRONMENT

United States

A mining engineer may be involved at the end of the mine life cycle when mine reclamation operations are planned and carried out. They also decide how to close a mine that has ceased operations to keep the public safe. Land reclamation is regulated for surface and underground mines according to the Surface Mining Control and Reclamation Act of 1977. The law creates as a part of the Department of Interior, the Bureau of Surface Mining (OSM). OSM states on their website, "OSM is charged with balancing the nation's need for continued domestic coal production with protection of the environment." The law requires that states set up their own Reclamation Departments and legislate laws related to reclamation for coal mining operations. The states may impose additional regulations and regulate other minerals in addition to coal for land reclamation.

ENVIRONMENTAL IMPACT OF MINING

Fig. Acid mine drainage in the Rio Tinto River.

The environmental impact of mining includes erosion, formation of sinkholes, loss of biodiversity, and contamination of soil,groundwater and surface water by chemicals from mining processes. In some cases, additional forest logging is done in the vicinity of mines to increase the available room for the storage of the created debris and soil. Besides creating environmental damage, the contamination resulting from leakage of chemicals also affect the health of the local population. Mining companies in some countries are required to follow environmental and rehabilitation codes, ensuring the area mined is returned to close to its original state. Some mining methods may have significant environmental and public health effects.

Erosion of exposed hillsides, mine dumps, tailings dams and resultant siltation of drainages, creeks and rivers can significantly impact the surrounding areas, a prime example being the giant Ok Tedi Mine in Papua New Guinea. In areas of wilderness mining may cause destruction and disturbance of ecosystems and habitats, and in areas of farming it may disturb or destroy productive grazing and croplands. In urbanised environments mining may produce noise pollution, dust pollution and visual pollution.

ISSUES

Water pollution

Fig. Acidic lake wastewater at the abandoned Northland Pyrite Mine inTemagami, Ontario, Canada.

Mining can have adverse effects on surrounding surface and ground water if protective measures are not taken. The result can be unnaturally high concentrations of some chemicals, such as arsenic, sulfuric acid, and mercury over a significant area of surface or subsurface. Runoff of mere soil or rock debris -although non-toxic- also devastates the surrounding vegetation. The dumping of the runoff in surface waters or in forests is the worst option here.

Submarine tailings disposal is regarded as a better option (if the soil is pumped to a great depth). Mere land storage and refilling of the mine after it has been depleted is even better, if no forests need to be cleared for the storage of the debris. There is potential for massive contamination of the area surrounding mines due to the various chemicals used in the mining process as well as the potentially damaging compounds and metals removed from the ground with the ore. Large amounts of water produced from mine drainage, mine cooling, aqueous extraction and other mining processes increases the potential for these chemicals to contaminate ground and surface water. In well-regulated mines, hydrologists and geologists take careful measurements of water and soil to exclude any type of water contamination that could be caused by the mine's operations. The reducing or eliminating of environmental degradation is enforced in modern American mining by federal and state law, by restricting operators to meet standards for protecting surface and ground water from contamination. This is best done through the use of non-toxic extraction processes as bioleaching. If the project site becomes nonetheless polluted, mitigation techniques such as acid mine drainage (AMD) need to be performed. The five principal technologies used to monitor and control water flow at mine sites are diversion systems, containment ponds, groundwater pumping systems, subsurface drainage systems, and subsurface barriers. In the case of AMD, contaminated water is generally pumped to a treatment facility that neutralizes the contaminants. A 2006 review of environmental impact statements found that "water quality predictions made after considering the effects of mitigations largely underestimated actual impacts to groundwater, seeps, and surface water".

Acid rock drainage

Heavy metals

Dissolution and transport of metals and heavy metals by run-off and ground water is another example of environmental problems with mining, such as the Britannia Mine, a former copper mine near Vancouver, British Columbia. Tar Creek, an abandoned mining area in Picher, Oklahoma that is now an Environmental Protection Agency superfundsite, also suffers from heavy metal contamination. Water in the mine containing dissolved heavy metals such as lead and cadmium leaked into local groundwater, contaminating it. Long-term storage of tailings and dust can lead to additional problems, as they can be easily blown off site by wind, as occurred at Scouriotissa, an abandoned copper mine in Cyprus.

Coal mining

Deforestation

With open cast mining the overburden, which may be covered in forest, must be removed before the mining can commence. Although the deforestation due to mining may be small compared to the total amount it may lead to species extinction if there is a high level of local endemism.

Salt Mining

Salt mining and salt dome collapsing in Assumption Parish, Louisiana caused the Bayou Corne sinkhole in 2012. As of August 2013, the sinkhole continues to expand.

- Tailings and spoil
- Tailings
- Slag heap
- Spoil Tip

MITIGATION

To ensure completion of reclamation, or restoring mine land for future use, many governments and regulatory authorities around the world require that mining companies post a bond to be held in escrow until productivity of reclaimed land has been convincingly demonstrated, although if cleanup procedures are more expensive than the size of the bond, the bond may simply be abandoned. Since 1978 the mining industry has reclaimed more than 2 million acres (8,000 km^2) of land in the United States alone. This reclaimed land has renewed vegetation and wildlife in previous mining lands and can even be used for farming and ranching.

3

Types of Mining and Techniques

SURFACE MINING

Fig. Coal strip mine in Wyoming

Surface mining, including strip mining, open-pit mining and mountaintop removal mining, is a broad category of mining in which soil and rock overlying the mineral deposit (the overburden) are removed. It is the opposite of underground mining, in which the overlying rock is left in place, and the mineral removed through shafts or tunnels.

Surface mining began in the mid-sixteenth century and is practiced throughout the world, although the majority of surface mining occurs in North America. It gained popularity throughout the 20th century, and is now the predominant form of mining in coal beds such as those in Appalachia and America's Midwest.

In most forms of surface mining, heavy equipment, such as earthmovers, first remove the overburden. Next, huge machines, such as dragline excavators or Bucket wheel excavators, extract the mineral.

TYPES[EDIT]

There are five main forms of surface mining, detailed below.

Strip mining

Fig. The Bagger 288 is a bucket-wheel excavator used in strip mining.

"Strip mining" is the practice of mining a seam of mineral, by first removing a long strip of overlying soil and rock (theoverburden). It is most commonly used to mine coal and lignite (brown coal). Strip mining is only practical when the ore body to be excavated is relatively near the surface. This type of mining uses some of the largest machines on earth, including bucket-wheel excavators which can move as much as 12,000 cubic meters of earth per hour. There are two forms of strip mining. The more common method is "area stripping", which is used on fairly flat terrain, to extract deposits over a large area. As each long strip is excavated, the overburden is placed in the excavation produced by the previous strip.

"Contour stripping" involves removing the overburden above the mineral seam near the outcrop in hilly terrain, where the mineral outcrop usually follows the contour of the land. Contour stripping is often followed by auger mining into the hillside, to remove more of the mineral. This method commonly leaves behind terraces in mountainsides.

Fig. Strip mining at Garzweiler, Germany. The lignite being extracted is at left, the removed overburden being placed at right. Note that it is a largely flat mine for a horizontal mineral.

Open-pit mining

Fig. The El Chino mine located near Silver City, New Mexico is an open-pit copper mine.

"Open-pit mining" refers to a method of extracting rock or minerals from the earth through their removal from an open pit or borrow. Although open-pit mining is sometimes mistakenly referred to as "strip mining", the two methods are different.

Mountaintop Removal

"Mountaintop removal mining" (MTR) is a form of coal mining that uses explosives to blast "overburden" off the top of some Appalachian mountains. Excess mining waste or "overburden" is dumped by large trucks into fills in nearby hollow or valley fills. MTR involves the mass restructuring of earth in order to reach the coal seam as deep as 400 feet (120 m) below the surface. Mountaintop removal replaces the original steep landscape with a much flatter topography, which may be suited to various different uses.

Economic development attempts on reclaimed mine sites include prisons such the Big Sandy Federal Penitentiary in Martin County, Kentucky, small town airports, golf courses such as Twisted Gun in Mingo County, West Virginia and Stonecrest Golf Course in Floyd County, Kentucky, as well as industrial scrubber sludge disposal sites, solid waste landfills, trailer parks, explosive manufacturers, and storage rental lockers. The technique has been used increasingly in recent years in the Appalachian coal fields of West Virginia, Kentucky, Virginia and Tennessee in the United States. The profound changes in topography and disturbance of pre-existing ecosystems have made mountaintop removal highly controversial.

Advocates of mountaintop removal point out that once the areas are reclaimed as mandated by law, the technique provides premium flat land suitable for many uses in a region where flat land is at a premium. They also maintain that the new growth on reclaimed mountaintop mined areas is better able to support populations of game animals. Critics contend that mountaintop removal is a disastrous practice that benefits a small number of corporations at the expense of local communities and the environment. AU.S. Environmental Protection Agency (EPA) environmental impact statement finds that streams near valley fills sometimes may contain higher levels of minerals in the water and decreased aquatic biodiversity. The statement also estimates that 724 miles (1,165 km) of Appalachian streams were buried by valley fills from 1985 to 2001.

Blasting at a mountaintop removal mine expels dust and fly-rock into the air, which can then disturb or settle onto private property nearby. This dust may contain sulfur compounds, which some claim corrode structures and tombstones and is a health hazard. Although MTR sites are required to be reclaimed after mining is complete, reclamation has traditionally focused on stabilizing rock and controlling erosion, but not always onreforesting the area. Quick-growing, non-native grasses, planted to quickly provide vegetation on a site, compete with tree seedlings, and trees have difficulty establishing root systems in compacted backfill. Consequently, biodiversity suffers in a region of the United States with numerous endemic species. Erosion also increases, which can intensify flooding. In the Eastern United States, the Appalachian Regional Reforestation Initiative works to promote the use of trees in mining reclamation.

Dredging

"Dredging" is a method often used to bring up underwater mineral deposits. Although dredging is usually employed to clear or enlarge waterways for boats, it can also recover significant amounts of underwater minerals relatively efficiently and cheaply. Used widely in oil refining.

Highwall Mining

Highwall mining is another form of surface mining that evolved from auger mining. In Highwall mining, the coal seam is penetrated by a continuous miner propelled by a hydraulic Pushbeam Transfer Mechanism (PTM). A typical cycle includes sumping (launch-pushing forward) and shearing (raising and lowering the cutterhead boom to cut the entire height of the coal seam). As the coal recovery cycle continues, the cutterhead is progressively launched into the coal seam for 19.72 feet (6.01 m). Then, the Pushbeam Transfer Mechanism (PTM) automatically inserts a 19.72-foot (6.01 m) long rectangular Pushbeam (Screw-Conveyor Segment) into the center section of the machine between the Powerhead and the cutterhead.

The Pushbeam system can penetrate nearly 1,000 feet (300 m) into the coal seam. One patented Highwall mining systems use augers enclosed inside the Pushbeam that prevent the mined coal from being contaminated by rock debris during the conveyance process. Using a video imaging and/or a gamma ray sensor and/or other Geo-Radar systems like a coal-rock interface detection sensor (CID), the operator can see ahead projection of the seam-rock interface and guide the continuous miner's progress. Highwall mining can produce thousands of tons of coal in contour-strip operations with narrow benches, previously mined areas, trench mine applications and steep-dip seams with controlled water-inflow pump system and/or a gas (inert) venting system. Recovery is much better than Augering, but the mapping of areas that have been developed by a Highwall miner are not mapped as rigorously as deep mined areas.

Very little soil is displaced in contrast with mountain top removal; however a large amount of capital is required to operate and own a Highwall miner. But then this Highwall mining system is the innovative roadmap future potential and stay or being better competitive in the area of environmental friendly non mountain-top (overburden) removal operated by only 4 crew members. Mapping of the outcrop as well as core hole data and samples taken during the bench making process are taken into account to best project the panels that the Highwall miner will cut. Obstacles that could be potentially damaged by subsidence and the natural contour of the Highwall are taken into account, and a surveyor points the Highwall miner in a line (Theoretical Survey Plot-Line) mostly perpendicular to the Highwall. Parallel lines represent the drive cut into the mountain (up to 1,000 feet (300 m) deep), without heading or corrective steering actuation on a navigation Azimuth during mining results in missing a portion of the coal seam and is a potential danger of cutting in pillars from previous mined drives due to horizontal drift (Roll) of the Pushbeam-Cuttermodule string. Recently Highwall miners have penetrated more than 1050 feet into the coal seam, and today's models are capable of going farther, with the support of gyro navigation and not limited anymore by the amount of cable stored on the machine.

The maximum depth would be determined by the stress of further penetration and associated specific-power draw, ("Torsion and Tension" in Screw-Transporters String) but today's optimized Screw-Transporters Conveying Embodiments (called: Pushbeams) with Visual Product Development and Flow Simulation Behaviour software "Discrete Element Modeling" (DEM) shows smart-drive extended penetrations are possible, evenso under steep inclined angles from horizontal to more than 30 degree downhole. In case of significant steep mining the new mining method phrase should be "Directional Mining", dry or wet, Dewatering is developed or Cutting & Dredging through Screw-Transporters are proactive in developing roadmap of the leading global Highwall mining company.

ENVIRONMENTAL AND HEALTH ISSUES

The large impact of surface mining on the topography, vegetation, and water resources has made it highly controversial. Surface mining is subject to state and federal reclamation requirements, but adequacy of the requirements is a constant source of contention. Unless reclaimed, surface mining can leave behind large areas of infertile waste rock, as 70% of material excavated is waste.

In the United States, the Surface Mining Control and Reclamation Act of 1977 mandates reclamation of surface coal mines. Reclamation for non-coal mines is regulated by state and local laws, which may vary widely.

Human health

The United Mine Workers of America has spoken against the use of human sewage sludge to reclaim surface mining sites in Appalachia. The UMWA launched its campaign against the use of sludge on mine sites in 1999 after eight UMWA workers became ill from exposure to Class B sludge spread near their workplace.

Environmental impact

According to a 2010 report in the journal *Science*, mountaintop mining has caused numerous environmental problems which mitigation practices have not successfully addressed. For example, valley fills frequently bury headwater streams causing permanent loss of ecosystems. In addition, the destruction of large tracts of deciduous forests has threatened several endangered species and led to a loss of biodiversity.

OPEN-PIT MINING

Fig. A rock quarry in Klunst, Germany

Fig. Sunrise Dam Gold Mine, Australia

Open-pit mining, open-cut mining or opencast mining is a surface mining technique of extracting rock orminerals from the earth by their removal from an open pit or borrow. This form of mining differs from extractive methods that require tunneling into the earth such as long wall mining. Open-pit mines are used when deposits of commercially useful minerals or rock are found near the surface; that is, where the overburden (surface material covering the valuable deposit) is relatively thin or the material of interest is structurally unsuitable for tunneling (as would be the case for sand, cinder, and gravel). For minerals that occur deep below the surface - where the overburden is thick or the mineral occurs as veins in hard rock - underground mining methods extract the valued material.

Open-pit mines that produce building materials and dimension stone are commonly referred to as quarries. Open-pit mines are typically enlarged until either the mineral resource is exhausted, or an increasing ratio of overburden to ore makes further mining uneconomic. When this occurs, the exhausted mines are sometimes converted to landfills for disposal of solid wastes. However, some form of water control is usually required to keep the mine pit from becoming a lake, if the mine is situated in a climate of considerable precipitation or if any layers of the pit forming the mine border productive aquifers.

EXTRACTION

Open-pit mines are dug on benches, which describe vertical levels of the hole. These benches are usually on four meter to sixty meter intervals, depending on the size of the machinery that is being used. Many quarries do not use benches, as they are usually shallow. Most walls of the pit are generally dug on an angle less than vertical, to prevent and minimize damage and danger from rock falls. This depends on how weathered the rocks are, and the type of rock, and also how many structural weaknesses occur within the rocks, such as a fault, shears, joints or foliations.

The walls are stepped. The inclined section of the wall is known as the batter, and the flat part of the step is known as the bench or berm. The steps in the walls help prevent rock falls continuing down the entire face of the wall. In some instances additional ground support is required and rock bolts, cable bolts and shotcrete are used. De-watering bores may be used to relieve water pressure by drilling horizontally into the wall, which is often enough to cause failures in the wall by itself. A haul road is usually situated at the side of the pit, forming a ramp up which trucks can drive, carrying ore and waste rock.

Waste rock is piled up at the surface, near the edge of the open pit. This is known as the waste dump. The waste dump is also tiered and stepped, to minimize degradation. Ore which has been processed is known as tailings, and is generally a slurry. This is pumped to a tailings dam or settling pond, where the water evaporates. Tailings dams can often be toxic due to the presence of unextracted sulfide minerals, some forms of toxic minerals in the gangue, and often cyanide which is used to treat gold ore via the cyanide leach process. This toxicity can harm the surrounding environment.

Rehabilitation

After mining finishes, the mine area must undergo rehabilitation. Waste dumps are contoured to flatten them out, to further stabilise them. If the ore contains sulfides it is usually covered with a layer of clay to prevent access of rain and oxygen from the air, which can oxidise the sulfides to produce sulfuric acid, a phenomenon known as acid mine drainage. This is then generally covered with soil, and vegetation is planted to help consolidate the material. Eventually this layer will erode, but it is generally hoped that the rate of leaching or acid will be slowed by the cover such that the environment can handle the load of acid and associated heavy metals.

There are no long term studies on the success of these covers due to the relatively short time in which large scale open pit mining has existed. It may take hundreds to thousands of years for some waste dumps to become "acid neutral" and stop leaching to the environment. The dumps are usually fenced off to prevent livestock denuding them of vegetation. The open pit is then surrounded with a fence, to prevent access, and it generally eventually fills up with ground water. In arid areas it may not fill due to deep groundwater levels.

Fig. Open-pit sulfur mining at Tarnobrzeg, Poland currently in land rehabilitation process

TYPICAL OPEN CUT GRADES

Gold is generally extracted in open-pit mines at 1 to 2 ppm (parts per million) but in certain cases, 0.75ppm gold is economical. This was achieved by bulk heap leaching at Alkane Minerals Ltd. Peak Hill mine in western New South Wales, near Dubbo, Australia.

Nickel, generally as laterite, is extracted via open-pit down to 0.2%. Copper is extracted at grades as low as 0.15% to 0.2%, generally in massive open-pit mines in Chile, where the size of the resources and favorable metallurgy allows economies of scale.

MOUNTAINTOP REMOVAL MINING

Fig. Mountaintop removal site

Mountaintop removal mining (MTR), also known as mountaintop mining (MTM), is a form of surface mining that involves the mining of the summit or summit ridge of a mountain. Coal seams are extracted from a mountain by removing the land, oroverburden, above the seams. The land may be dumped back on the ridge and compacted to reflect the approximate original contour of the mountain. It is the predominant method of coal mining in the Appalachian Mountains in the eastern United States.Explosives are used to remove up to 400 vertical feet (120 m) of mountain to expose underlying coal seams. Excess rock and soil laden with toxic mining byproducts are often dumped into nearby valleys, in what are called "holler fills" or "valley fills." Less expensive to execute and requiring fewer employees, mountaintop removal mining began in Appalachia in the 1970s as an extension of conventional strip mining techniques. It is primarily occurring in Kentucky, West Virginia, Virginia, and Tennessee

Peer-reviewed studies show that mountaintop mining has serious environmental impacts, including loss of biodiversity and toxification of watersheds, that mitigation practices cannot successfully address. There are also adverse human health impacts which result from contact with affected streams or exposure to airborne toxins and dust. According to 21 scientific studies there has been major effects on the population in the Appalachia where MTM takes place including over 50% higher cancer rates, 42% higher birth defect rates, and $75 billion a year in public health costs from pollution. Mountaintop removal mining (MTR), also known as mountaintop mining (MTM), is a form of surface mining that involves the topographical alteration and/or removal of asummit, hill, or ridge to access buried coal seams.

The MTR process involves the removal of coal seams by first fully removing the overburden laying atop them, exposing the seams from above. This method differs from more traditional underground mining, where typically a narrow shaft is dug which allows miners to collect seams using various underground methods, while leaving the vast majority of the overburden undisturbed. The overburden waste resulting from MTR is either placed back on the ridge, attempting to reflect the approximate original contour of the mountain, and/or it is moved into neighboring valleys. Excess rock and soil containing mining byproducts are disposed into nearby valleys, in what are called "holler fills" or "valley fills."

MTR in the United States is most often associated with the extraction of coal in the Appalachian Mountains, where the United States Environmental Protection Agency (EPA) estimates that 2,200 square miles (5,700 km) of Appalachian forests will be cleared for MTR sites by the year 2012. Sites range from Ohio to Virginia. It occurs most commonly in West Virginia and Eastern Kentucky, the top two coal-producing states in Appalachia, with each state using approximately 1,000 tonnes of explosives per day for surface mining. At current rates, MTR in the U.S. will mine over 1.4 million acres (5,700 km^2)

by 2010, an amount of land area that exceeds that of the state of Delaware. Mountaintop removal has been practiced since the 1960s. Increased demand for coal in the United States, sparked by the 1973 and 1979 petroleum crises, created incentives for a more economical form of coal mining than the traditional underground mining methods involving hundreds of workers, triggering the first widespread use of MTR.

Its prevalence expanded further in the 1990s to retrieve relatively low-sulfur coal, a cleaner-burning form, which became desirable as a result of amendments to the U.S. Clean Air Act that tightened emissions limits on high-sulfur coal processing.

PROCESS

Land is deforested prior to mining operations and the resultant lumber is either sold or burned. According to the Surface Mining Control and Reclamation Act of 1977 (SMCRA), the topsoil is supposed to be removed and set aside for later reclamation.However, coal companies are often granted waivers and instead reclaim the mountain with "topsoil substitute." The waivers are granted if adequate amounts of topsoil are not naturally present on the rocky ridge top.

Once the area is cleared, miners use explosives to blast away the overburden, the rock and subsoil, to expose coal seams beneath. The overburden is then moved by various mechanical means to areas of the ridge previously mined. These areas are the most economical area of storage as they are located close to the active pit of exposed coal. If the ridge topography is too steep to adequately handle the amount of spoil produced then additional storage is used in a nearby valley or hollow, creating what is known as a *valley fill* or *hollow fill*. Any streams in a valley are buried by the overburden. A front-end loader or excavator then removes the coal, where it is transported to a processing plant.

Once coal removal is completed, the mining operators back stack overburden from the next area to be mined into the now empty pit. After backstacking and grading of overburden has been completed, topsoil (or a topsoil substitute) is layered over the overburden layer. Next, grass seed is spread in a mixture of seed, fertilizer, and mulch made from recycled newspaper. Depending on surface land owner wishes the land will then be further reclaimed by adding trees if the pre-approved post-mining land use is forest land or wildlife habitat. If the land owner has requested other post-mining land uses the land can be reclaimed to be used as pasture land, economic development or other uses specified in SMCRA.

Because coal usually exists in multiple geologically stratified seams, miners can often repeat the blasting process to mine over a dozen seams on a single mountain, increasing the mine depth each time. This can result in a vertical descent of hundreds of extra feet into the earth.

ECONOMICS

Almost half of the electricity generated in the United States is produced by coal-fired power plants. MTR accounted for less than 5% of U.S. coal production as of 2001. In some regions, however, the percentage is higher, for example MTR provided 30% of the coal mined in West Virginia in 2006. Historically in the U.S. the prevalent method of coal acquisition was underground mining which is very labor-intensive.

In MTR, through the use of explosives and large machinery, more than two and a half times as much coal can be extracted per worker per hour than in traditional underground mines, thus greatly reducing the need for workers. In Kentucky, for example, the number of workers has declined over 60% from 1979 to 2006 (from 47,190 to 17,959 workers). The industry overall lost approximately 10,000 jobs from 1990 to 1997, as MTR and other more mechanized underground mining methods became more widely used. The coal industry asserts that surface mining techniques, such as mountaintop removal, are safer for miners than sending miners underground.

Proponents argue that in certain geologic areas, MTR and similar forms of surface mining allow the only access to thin seams of coal that traditional underground mining would not be able to mine. MTR is sometimes the most cost-effective method of extracting coal. Several studies of the impact of restrictions to mountaintop removal were authored in 2000 through 2005. Studies by Mark L. Burton, Michael J. Hicks and Cal Kent identified significant state level tax losses attributable to lower levels of mining (notably the studies did not examine potential environmental costs, which the authors acknowledge may outweigh commercial benefits).

LEGISLATION IN THE UNITED STATES

In the United States, MTR is allowed by section 515(c)(1) of the Surface Mining Control and Reclamation Act of 1977. Although most coal mining sites must be reclaimed to the land's pre-mining contour and use, regulatory agencies can issue waivers to allow MTR. In such cases, SMCRA dictates that reclamation must create "a level plateau or a gently rolling contour with no highwalls remaining."

Permits must be obtained to deposit valley fill into streams. On four occasions, federal courts have ruled that the US Army Corps of Engineers violated the Clean Water Act by issuing such permits. Massey Energy Company is currently appealing a 2007 ruling, but has been allowed to continue mining in the meantime because "most of the substantial harm has already occurred," according to the judge.

The Bush administration appealed one of these rulings in 2001 because the Act had not explicitly defined "fill material" that could legally be placed in a waterway. The EPA and Army Corps of Engineers changed a rule to include mining debris in the definition of fill material, and the ruling was overturned. On December 2, 2008, the Bush Administration made a rule change to remove the Stream Buffer Zone protection provision from SMCRA allowing coal companies to place mining waste rock and dirt directly into headwater waterways.

A federal judge has also ruled that using settling ponds to remove mining waste from streams violates the Clean Water Act. He also declared that the Army Corps of Engineers has no authority to issue permits allowing discharge of pollutants into such in-stream settling ponds, which are often built just below valley fills. On January 15, 2008, the environmental advocacy group Center for Biological Diversity petitioned the United States Fish and Wildlife Service (FWS) to end a policy that waives detailed federal Endangered Species Act reviews for new mining permits. The current policy states that MTR can never damage endangered species or their habitat as long as mining operators comply with federal surface mining law, despite the complexities of species and ecosystems.

Since 1996, this policy has exempted many strip mines from being subject to permit-specific reviews of impact on individual endangered species. Because of the 1996 Biological Opinion by FWS making case-by-case formal reviews unnecessary, the Interior's Office of Surface Mining and state regulators require mining companies to hire a government-approved contractor to conduct their own surveys for any potential endangered species. The surveys require approval from state and federal biologists, who provide informal guidance on how to minimize mines' potential effects to species. While the agencies have the option to ask for formal endangered species consultations during that process, they do so very rarely. On May 25, 2008, North Carolina State Representative Pricey Harrison introduced a bill to ban the use of mountaintop removal coal from coal-fired power plants within North Carolina. This proposed legislation would have been the only legislation of its kind in the United States; however, the bill was defeated.

ENVIRONMENTAL AND HEALTH IMPACTS

Critics contend that MTR is a destructive and unsustainable practice that benefits a small number of corporations at the expense oflocal communities and the environment. Though the main issue has been over the physical alteration of the landscape, opponents to the practice have also criticized MTR for the damage done to the environment by massive transport trucks, and the environmental damage done by the burning of coal for power. Blasting at MTR sites also expels dust and fly-rock into the air, which can disturb or settle onto private property nearby. This dust may contain sulfur compounds, which corrodes structures and is a health hazard.

A January 2010 report in the journal *Science* reviews current peer-reviewed studies and water quality data and explores the consequences of mountaintop mining. It concludes that mountaintop mining has serious environmental impacts that mitigation practices cannot successfully address. For example, the extensive tracts of deciduous forests destroyed by mountaintop mining support several endangered species and some of the highest biodiversity in North America. There is a particular problem with burial of headwater streams by valley fills which causes permanent loss of ecosystems that play critical roles in ecological processes. In addition, increases in metal ions, pH, electrical conductivity, total dissolved solids due to elevated concentrations of sulfate are closely linked to the extent of mining in West Virginia watersheds. Declines in stream biodiversity have been linked to the level of mining disturbance in West Virginia watersheds.

Published studies also show a high potential for human health impacts. These may result from contact with streams or exposure to airborne toxins and dust. Adult hospitalization for chronic pulmonary disorders and hypertension are elevated as a result of county-level coal production. Rates of mortality, lung cancer, as well as chronic heart, lung and kidney disease are also increased. A 2011 study found that counties in and near mountaintop mining areas had higher rates of birth defects for five out of six types of birth defects, including circulatory/respiratory, musculoskeletal, central nervous system, gastrointestinal, and urogenital defects. These defect rates were more pronounced in the most recent period studied, suggesting the health effects of mountaintop mining-related air and water contamination may be cumulative. Another 2011 study found "the odds for reporting cancer were twice as high in the mountaintop mining environment compared to the non mining environment in ways not explained by age, sex, smoking, occupational exposure, or family cancer history."

A United States Environmental Protection Agency (EPA) environmental impact statement finds that streams near some valley fills from mountaintop removal contain higher levels of minerals in the water and decreased aquatic biodiversity. The statement also estimates that 724 miles (1,165 km) of Appalachian streams were buried by valley fills between 1985 to 2001. On September 28, 2010, the U.S. Environmental Protection Agency's (EPA) independent Science Advisory Board (SAB) released their first draft review of EPA's research into the water quality impacts of valley fills associated with mountaintop mining, agreeing with EPA's conclusion that valley fills are associated with increased levels of conductivity threatening aquatic life in surface waters.

Although U.S. mountaintop removal sites by law must be reclaimed after mining is complete, reclamation has traditionally focused on stabilizing rock formations and controlling for erosion, and not on the reforestation of the affected area. Fast-growing, non-native flora such as *Lespedeza cuneata*, planted to quickly provide vegetation on a site, compete with tree seedlings, and trees have difficulty establishing root systems in compacted backfill. Consequently, biodiversity suffers in a region of the United States with numerous endemic species. In addition, reintroduced elk (*Cervus canadensis*) on mountaintop removal sites in Kentucky are eating tree seedlings. Advocates of MTR claim that once the areas are reclaimed as mandated by law, the area can provide flat land suitable for many uses in a region where flat land is at a premium. They also maintain that the new growth on reclaimed mountaintop mined areas is better suited to support populations of game animals. While some of the land is able to be turned into grassland which game animals can live in, the amount of grassland is minimal. The land does not retake the form it had before the MTR. As stated in the book Bringing Down the Mountains: "Some of the main problems associated with MTR include soil depletion, sedimentation, low success rate of tree regrowth, lack of successful revegetation, displacement of native wildlife, and burial of streams." The ecological benefits after MTR are far below the level of the original land.

Books and films

Dr. Shirley Stewart Burns, a West Virginia coalfield native, wrote the first academic work on mountaintop removal, titled *Bringing Down The Mountains* (2007), which is loosely based on her internationally award-winning 2005 Ph.D. dissertation of the same name. Dr. Burns was also a co-editor with Kentucky author Silas House and filmmaker Mari-Lynn Evans, for the 2009 book, "Coal Country", which was a companion book for the nationally recognized feature-length film of the same name. Dr. Rebecca Scott, another native West Virginian, examined the sociological relationship of identity and natural resource extraction in central Appalachia in her 2010 book, "Removing Mountains". Cultural historian Jeff Biggers has also published *The United States of Appalachia* examined the cultural and human costs of mountaintop removal.

Additionally, many personal interest stories of coalfield residents have been written, including *Lost Mountain* (2007) by Erik Reese and *Moving Mountains: How One Woman and Her Community Won Justice From Big Coal* (2007) by Penny Loeb. In April 2005, a group of Kentucky writers traveled together to see the devastation from mountaintop removal mining, and Wind Publishing produced the resulting collection of poems, essays and photographs, co-edited by Kristin Johannesen, Bobbie Ann Mason and Mary Ann Taylor-Hall — *Missing Mountains: We went to the mountaintop, but it wasn't there.* In 2007, Ann Pancake released the novel *Strange As This Weather Has Been,* the first major fiction work about the subject. Mountaintop removal is a major plot element of the 2010 best-selling novel *Freedom* by Jonathan Franzen, wherein a major character helps to secure land for surface mining with the promise that it will be restored and turned into a nature reserve.

In 2006, Catherine Pancake released the first comprehensive feature-length documentary on mountaintop removal, *Black Diamonds: Mountaintop Removal and the Search for Coalfield Justice,* a selection in the Documentary Fortnight at the Museum of Modern Art. The film features Julia Bonds who won the 2003 Goldman Prize. A 2007 documentary, *Mountain Top Removal,* focuses on Mountain Justice Summer activists, coal field residents, and coal industry officials. On April 18, 2008, the film received the Reel Current award selected and presented by Al Gore at the Nashville Film Festival. Another feature documentary, titled *Burning the Future: Coal in America,* was awarded the International Documentary Association's 2008 Pare Lorentz award for Best Documentary. In 2011, the film *The Last Mountain* directed by Bill Haney detailed the effects on the land and people living near mountaintop removal and coal burning sites. Maria Gunnoe, the 2009 Goldman Environmental Prize winner, Robert F. Kennedy, Jr. and others present the devastation, confront the politicians and corporate interests, and offer wind power as one solution for Coal River Mountain, WV.

In 2013, the autoethnographic documentary film *Goodbye Gauley Mountain: An Ecosexual Love Story,* by Beth Stephens with Annie Sprinkle, raises awareness on the issue of MTR in West Virginia by bringing together environmental activism, performance art and queer activism against the issue of MTR. Stephens says: "My hope for this film, is that in addition to it being a compelling story, it will inspire and raise awareness in groups of people not normally associated with the environmental movement, and especially in GLBTQI communities. (We'd like to see an E added for ecosexual!) There are relatively few films about environmental issues that feature out queers."

UNDERGROUND MINING (HARD ROCK)

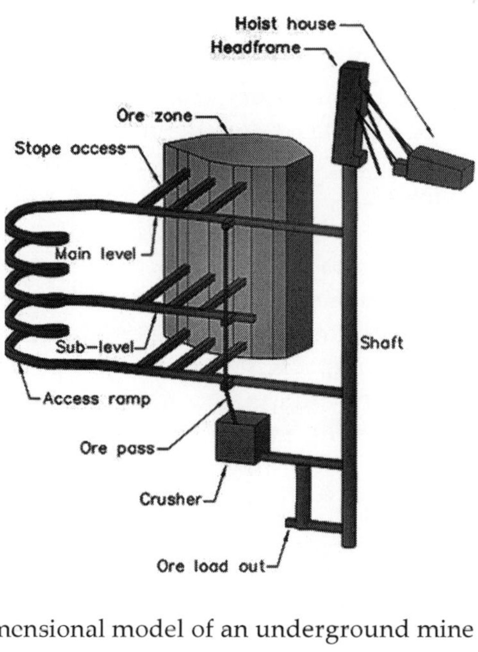

Fig. A three dimensional model of an underground mine with shaft access

Underground hard rock mining refers to various underground mining techniques used to excavate *hard* minerals, mainly those containing metals such as ore containing gold, silver, iron, copper, zinc, nickel, tin and lead, but also involves using the same techniques for excavating ores of gems such as diamonds. In contrast soft rock mining refers to excavation of softer minerals such as salt, coal, or oil sands.

MINE ACCESS

Underground access

Accessing underground ore can be achieved via a decline (ramp), inclined vertical shaft or adit.

Fig. Decline portal atWiluna Gold Mine

Declines can be a spiral tunnel which circles either the flank of the deposit or circles around the deposit. The decline begins with a box cut, which is the portal to the surface. Depending on the amount of overburden and quality of bedrock, a galvanized steel culvert may be required for safety purposes. They may also be started into the wall of an open cut mine.

Shafts are vertical excavations sunk adjacent to an ore body. Shafts are sunk for ore bodies where haulage to surface via truck is not economical. Shaft haulage is more economical than truck haulage at depth, and a mine may have both a decline and a ramp. Adits are horizontal excavations into the side of a hill or mountain. They are used for horizontal or near-horizontal ore bodies where there is no need for a ramp or shaft.

Declines are often started from the side of the high wall of an open cut mine when the ore body is of a payable grade sufficient to support an underground mining operation but the strip ratio has become too great to support open cast extraction methods. They are also often built and maintained as an emergency safety access from the underground workings and a means of moving large equipment to the workings.

Ore access

Levels are excavated horizontally off the decline or shaft to access the ore body. Stopes are then excavated perpendicular (or near perpendicular) to the level into the ore.

DEVELOPMENT MINING VS. PRODUCTION MINING

There are two principal phases of underground mining: development mining and production mining. Development mining is composed of excavation almost entirely in (non-valuable) waste rock in order to gain access to the orebody. There are six steps in development mining: remove previously blasted material (muck out round), Scaling (removing any unstable slabs of rock hanging from the roof and sidewalls to protect workers and equipment from damage), installing support or/and reinforcement, drill face rock, load explosives, and blast explosives.

Production mining is further broken down into two methods, long hole and short hole. Short hole mining is similar to development mining, except that it occurs in ore. There are several different methods of long hole mining. Typically long hole mining requires two excavations within the ore at different elevations below surface, (15 m – 30 m apart). Holes are drilled between the two excavations and loaded with explosives. The holes are blasted and the ore is removed from the bottom excavation.

VENTILATION

Fig. Door for directing ventilation in an old lead mine.
The ore hopper at the front is not part of the ventilation.

One of the most important aspects of underground hard rock mining is ventilation. Ventilation is required to clear toxic fumes from blasting and removing exhaust fumes from diesel equipment. In deep hot mines ventilation is also required for cooling the workplace for miners. Ventilation raises are excavated to provide ventilation for the workplaces, and can be modified for use as emergency escape routes. The primary sources of heat in underground hard rock mines are virgin rock temperature, machinery, auto compression, and fissure water. Other small contributing factors are human body heat and blasting.

GROUND SUPPORT

Some means of support is required in order to maintain the stability of the openings that are excavated. This support comes in two forms, local support and area support.

Area ground support

Area ground support is used to prevent major ground failure. Holes are drilled into the back (ceiling) and walls and a long steel rod (or rock bolt) is installed to hold the ground together. There are three categories of rock bolt, differentiated by how they engage the host rock. They are:

Mechanical bolts

Point anchor bolts (or expansion shell bolts) are a common style of area ground support. A point anchor bolt is a metal bar between 20 mm – 25 mm in diameter, and between 1 m – 4 m long (the size is determined by the mine's engineering department). There is an expansion shell at the end of the bolt which is inserted into the hole. As the bolt is tightened by the installation drill the expansion shell expands and the bolt tightens holding the rock together. Mechanical bolts are considered temporary support as their lifespan is reduced by corrosion as they are not grouted.

Grouted bolts

Resin grouted rebar is used in areas which require more support than a point anchor bolt can give. The rebar used is of similar size as a point anchor bolt but does not have an expansion shell. Once the hole for the rebar is drilled, cartridges of polyester resin are installed in the hole. The rebar bolt is installed after the resin and spun by the installation drill. This opens the resin cartridge and mixes it. Once the resin hardens the drill spinning tightens the rebar bolt holding the rock together. Resin grouted rebar is considered a permanent ground support with a lifespan of 20–30 years.

Cable bolts are used to bind large masses of rock in the hanging wall and around large excavations. Cable bolts are much larger than standard rock bolts and rebar, usually between 10–25 metres long. Cable bolts are grouted with a cement grout.

Friction bolts

- Friction stabilizer (frequently called by the genericized trademark

Split Set) are much easier to install than mechanical bolts or grouted bolts. The bolt is hammered into the drill hole, which has a smaller diameter than the bolt. Pressure from the bolt on the wall holds the rock together. Friction stabilizers are particularly susceptible to corrosion and rust from water unless they are grouted. Once grouted the friction increases by a factor of 3-4.

- Swellex is similar to Friction stabilizers, except the bolt diameter is smaller than the hole diameter. High pressure water is injected into the bolt to expand the bolt diameter to hold the rock together. Like the friction stabilizer, swellex is poorly protected from corrosion and rust.

Local ground support

Local ground support is used to prevent smaller rocks from falling from the backs and ribs. Not all excavations require local ground support.

Welded Wire Mesh is a metal screen with 10 cm x 10 cm (4 inch) openings. It is held to the backs using point anchor bolts or resin grouted rebar.

Shotcrete is fibre reinforced spray on concrete which coats the backs and ribs preventing smaller rocks from falling. Shotcrete thickness can be between 50 mm – 100 mm.

Latex Membranes can be sprayed on the backs and ribs similar to shotcrete, but in smaller amounts.

STOPE AND RETREAT VS. STOPE AND FILL

Stope and retreat

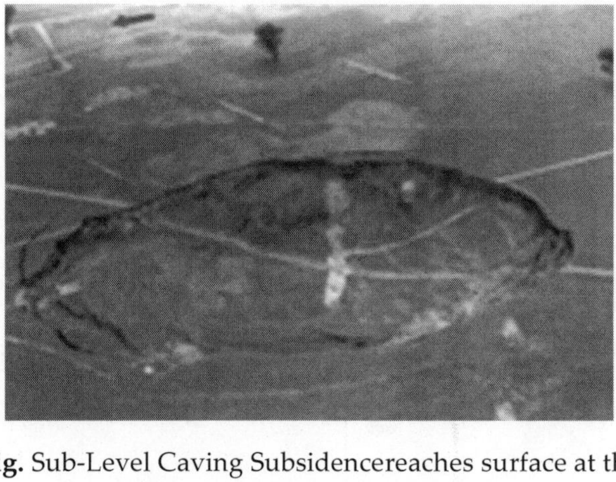

Fig. Sub-Level Caving Subsidencereaches surface at the Ridgeway underground mine.

Using this method, mining is planned to extract rock from the stopes without filling the voids; this allows the wall rocks to cave in to the extracted stope after all the ore has been removed. The stope is then sealed to prevent access.

Stope and fill

Where large bulk ore bodies are to be mined at great depth, or where leaving pillars of ore is uneconomical, the open stope is filled with backfill, which can be a cement and rock mixture, a cement and sand mixture or a cement and tailings mixture. This method is popular as the refilled stopes provide support for the adjacent stopes, allowing total extraction of economic resources.

MINING METHODS

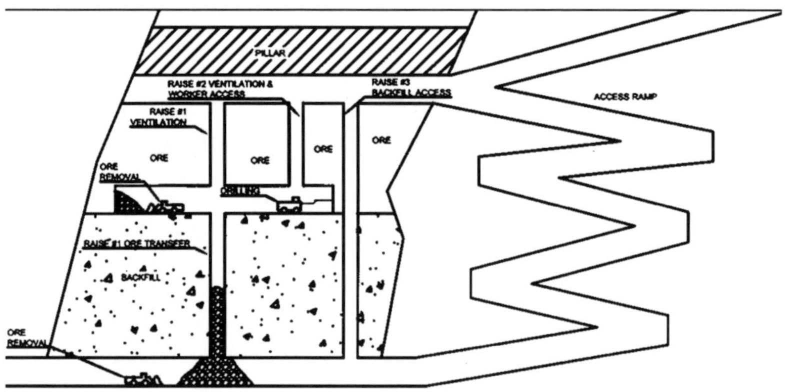

Fig. Schematic diagram of Cut and Fill mining

The mining method selected is determined by the size, shape, orientation and type of orebody to be mined. The orebody can be narrow vein such as a gold mine in the Witswaterand, the orebody can be massive similar to the Olympic Dam, South Australia orCadia-Ridgeway Mine, New South Wales. The width or size of the orebody is determined by the grade as well as the distribution of the ore. The dip of the orebody also has an influence on the mining method for example a narrow horizontal vein orebody will be mined by room and pillar or a longwall method whereas a vertical narrow vein orebody will be mined by an open stoping or cut and fill method. Futher consideration is needed for the strength of the ore as well as the surrounding rock. An orebody hosted in strong self supporting rock may be mined by an open stoping method and an orebody hosted in poor rock may need to be mined by a cut and fill method where the void is continuously filled as the ore is removed.

Selective mining methods

Cut and fill mining is a method of short hole mining used in steeply dipping or irregular ore zones, in particular where the hanging wall limits the use of long hole methods. The ore is mined in horizontal or slightly inclined slices, and then filled with waste rock, sand or tailings. Either fill option may be consolidated with concrete, or leftunconsolidated. Cut and fill mining is an expensive but selective method, with low ore loss and dilution.

Drift and fill is similar to cut and fill, except it is used in ore zones which are wider than the method of drifting will allow to be mined. In this case the first drift is developed in the ore, and is backfilled using consolidated fill. The second drift is driven adjacent to the first drift. This carries on until the ore zone is mined out to its full width, at which time the second cut is started atop of the first cut.

Shrinkage stoping is a short hole mining method which is suitable for steeply dipping orebodies. The method is similar to cut and fill mining with the exception that after being blasted, broken ore is left in the stope where it is used to support the surrounding rock and as a platform from which to work. Only enough ore is removed from the stope to allow for drilling and blasting the next slice. The stope is emptied when all of the ore has been blasted. Although it is very selective and allows for low dilution, since the most of the ore stays in the stope until mining is completed there is a delayed return on capital investments.

- Room and pillar mining: Room and pillar mining is commonly done in flat or gently dipping bedded ore bodies. Pillars are left in place in a regular pattern while the rooms are mined out. In many room and pillar mines, the pillars are taken out starting at the farthest point from the stope access, allowing the roof to collapse and fill in the stope. This allows for greater recovery as less ore is left behind in pillars.

Bulk mining methods

- Block caving is used to mine massive steeply dipping orebodies (typically low grade) with high friability. An undercut with haulage access is driven under the orebody, with "drawbells" excavated between the top of the haulage level and the bottom of the undercut. The drawbells serve as a place for caving rock to fall into. The orebody is drilled and blasted above the undercut, and the ore is removed via the haulage access. Due to the friability of the orebody the ore above the first blast caves and falls into the drawbells. As ore is removed from the drawbells the orebody caves in providing a steady stream of ore. If caving stops and removal of ore from the drawbells continues, a large void may form, resulting in the potential for a sudden and massive collapse and potentially catastrophic windblast throughout the mine. Where caving does continue, the ground surface may collapse into a surface depression such as those at the Climax and Henderson molybdenum mines in Colorado. Such a configuration is one of several to which miners apply the term "glory hole".

Orebodies that do not cave readily are sometimes preconditioned by hydraulic fracturing, blasting, or by a combination of both. Hydraulic fracturing has been applied to preconditioning strong roof rock over coal longwall panels, and to inducing caving in both coal and hard rock mines.

ORE REMOVAL

In mines which use rubber tired equipment for coarse ore removal, the ore (or "muck") is removed from the stope (referred to as "mucked out" or "bogged") using center articulated vehicles (referred to as boggers or LHD [short for Load, Haul, Dump]). These pieces of equipment may operate using diesel engines or electric motors, and resemble a low-profile front end loader.

The ore is then dumped into a truck to be hauled to the surface (in shallower mines). In deeper mines the ore is dumped down an ore pass (a vertical or near vertical excavation) where it falls to a collection level. On the collection level, it may receive primary crushing via jaw or cone crusher, or via a rockbreaker. The ore is then moved byconveyor belts, trucks or occasionally trains to the shaft to be hoisted to the surface in buckets or skips and emptied into bins beneath the surface headframe for transport to the mill.

In some cases the underground primary crusher feeds an inclined conveyor belt which delivers ore via an incline shaft direct to the surface. The ore is fed down ore passes, with mining equipment accessing the ore body via a decline from surface.

DEEPEST MINES

- The deepest mines in the world are the TauTona (Western Deep Levels) and Savuka gold mines in the Witwatersrand region of South Africa, which are currently working at depths exceeding 3,900 m (12,800 ft). There are plans to extend Mponeng mine, a sister mine to TauTona, down to 4,500 m (14,800 ft) in the coming years.
- The deepest hard rock mine in North America is Agnico-Eagle's LaRonde mine, which mines gold, zinc, copper and silver ores roughly 45 km (28 mi) east of Rouyn-Noranda in Cadillac, Quebec. LaRonde's Penna shaft (#3 shaft) is believed to be the deepest single lift shaft in the Western Hemisphere. The new #4 shaft bottoms out at over 3,000 m (9,800 ft) down. Their LaRonde mine expansion sees open stopes down to a depth of over 3,000 m (9,800 ft), the deepest longhole open stopes in the world.
- The deepest hard rock mines in Australia are the copper and zinc lead mines in Mount Isa, Queensland at 1,800 m (5,900 ft).
- The deepest platinum-palladium mines in the world are on the Merensky Reef, in South Africa, with a resource of 203 million Troy

ounces, currently worked to approximately 2,200 m (7,200 ft) depth.

- The harshest conditions for hard rock mining are in the Witwatersrand area of South Africa, where workers toil in temperatures of up to 45 °C (113 °F). However, massive refrigeration plants are used to bring the air temperature down to around 28 °C (82 °F).

UNDERGROUND MINING (SOFT ROCK)

Underground mining (soft rock) refers to a group of underground mining techniques used to extract coal, oil shale and other minerals or geological materials from sedimentary ("soft") rocks. Because deposits in sedimentary rocks are commonly layered and relatively less hard, the mining methods used differ from those used to mine deposits in igneous or metamorphic rocks (*see Underground mining (hard rock)*). Underground mining techniques also differ greatly from those of surface mining.

METHODS

- Longwall mining – A set of longwall mining equipment consists of a coal shearer mounted on conveyor operating underneath a series of self-advancing hydraulic roof supports. Almost the entire process can be automated. Longwall mining machines are typically 150–250 metres in width and 1.5 to 3 metres high. Longwall miners extract "panels" - rectangular blocks of coal as wide as the face the equipment is installed in, and as long as several kilometres. Powerful mechanical coal cutters (shearers) cut coal from the face, which falls onto an armoured face conveyor for removal. Longwalls can advance into an area of coal, or more commonly, retreat back between development tunnels (called "gateroads") As a longwall miner retreats back along a panel, the roof behind the supports is allowed to collapse in a planned and controlled manner.
- Room-and-pillar mining or continuous mining – Room and pillar mining is commonly done in flat or gently dipping bedded ores. Pillars are left in place in a regular pattern while the rooms are mined out. In many room and pillar mines, the pillars are taken out, starting at the farthest point from the mine haulage exit, retreating, and letting the roof come down upon the floor. Room and pillar methods are well adapted to mechanization, and are used in deposits such as coal, potash, phosphate, salt, oil shale, and bedded uranium ores.
- Blast mining – An older practice of coal mining that uses explosives such as dynamite to break up the coal seam, after which the coal is gathered and loaded onto shuttle cars or conveyors for removal to a central loading area. This process consists of a series of operations

that begins with "cutting" the coalbed so it will break easily when blasted with explosives. This type of mining accounts for less than 5% of total underground production in the U.S. today.

- Shortwall mining – A coal mining method that accounts for less than 1% of deep coal production, shortwall involves the use of a continuous mining machine with moveable roof supports, similar to longwall. The continuous miner shears coal panels 150–200 feet wide and more than a half-mile long, depending on other things like the strata of the Earth and the transverse waves.
- Coal skimming – While no longer in general use, because of the massive amount of water needed and environmental damage there of, in the late 1930s DuPontdeveloped a method that was much faster and less labour intensive than previous methods to separate the lighter coal from the mining refuse (*e.g.* slate) called "coal skimming" or the "sink and float method".

Mine shorthand

The number sign, or hash sign (#) is often used as shorthand to denote shaft or seam, as in 4# (4 shaft or seam depending on context).

DRIFT MINING

Drift mining is either the mining of an ore deposit by underground methods, or the working of coal seams accessed by adits driven into the surface outcrop of the coal bed.Drift is a more general mining term, meaning a near-horizontal passageway in a mine, following the bed (of coal, for instance) or vein of ore. A drift may or may not intersect the ground surface. This kind of mining is done when the rock or mineral is on the side of a hill.

Fig. Drift mine entry in West Virginia, 1908. Photo by Lewis Hine.

Nome, Alaska

Drift mining methods were used extensively to mine placer deposits during the early years of the Nome mining district. During summer, surface deposits could be worked, but some placer deposits were buried too deeply for surface placering.

In addition, water to wash the gold from the placers was not available in the winter. Many miners tunnelled into deep placer deposits, bringing out the high-grade gravels to be washed at the spring thaw. Most of the ground in Nome is permafrost. By drift mining, miners were able to recover much of the gold buried under the permafrost. Gold at Nome was concentrated in three ancient beach lines, now inshore, above sea level, and buried under roughly fifty feet of permafrost overlain by two feet of tundra. Gold was usually found on top of "false bedrock," a layer of clay that occurred at the base of the beach or stream deposit.

The miners initially sank shafts to prospect for the pay streaks. This was done by building a fire atop the permafrost, and, each day as it melted, shovelling away the mud. The process would be continued down to either a pay streak or bedrock. When the gold was found, the drift mining began. Once they had found a pay streak, miners would tunnel horizontally from the bottom of their prospect shaft, and follow the gold along the surface of the bedrock.

The tunnels, because the ground was frozen, would not cave in. Miners discovered old underground beach and river gravels rich with gold. Around 1900 the population of Nome was more than twenty thousand, many of them drift miners. Nome's gold fields, appearing untouched from the surface, are honeycombed with tunnels left by the gold rush drift miners. Today's miners, prospecting with modern drilling equipment, sometimes hit old drifts this was and is a mining technique copied from the Welsh coal miners of south Wales. This was more effective then using bell pits.

SHAFT MINING

Shaft mining or shaft sinking refers to the method of excavating a vertical or near-vertical tunnel from the top down, where there is initially no access to the bottom. Shaft sinking refers to the general activity, however shallow shafts, typically sunk for civil engineering projects differ greatly in execution method from deep shafts, typically sunk for mining projects.

When the top of the excavation is the ground surface, it is referred to as a shaft; when the top of the excavation is underground, it is called a winze or a sub-shaft. Small shafts may be excavated upwards from within an existing mine as long as there is access at the bottom, in which case they are called Raises. A shaft may be either vertical or inclined (between 45 and 90 degrees to the horizontal), although most modern mine shafts are vertical.

Fig. Abandoned mine shafts in Marl, Germany.

Shaft Sinking is one of the most difficult of all development methods: restricted space, gravity, groundwater and specialized procedures make the task quite formidable Historically mine shaft sinking has been among the most dangerous of all mining occupations and the preserve of mining contractors. Today shaft sinking contractors are concentrated in Canada and South Africa.

PARTS OF A MINE SHAFT

The most visible feature of a mine shaft is the headframe (or winding tower, poppet head or pit head) which stands above the shaft. Depending on the type of hoist used the top of the headframe will either house a hoist motor or a sheave wheel (with the hoist motor mounted on the ground). The headframe will also contain bins for storing ore being transferred to the processing facility.

At ground level beneath and around the headframe is the Shaft Collar (also called the Bank or Deck), which provides the foundation necessary to support the weight of the headframe and provides a means for men, materials and services to enter and exit the shaft. Collars are usually massive reinforced concrete structures with more than one level. If the shaft is used for mine ventilation, a plenum space or casing is incorporated into the collar to ensure the proper flow of air into and out of the mine. Beneath the Collar the part of the shaft which continues into the ground is called the Shaft Barrel.

At locations where the Shaft Barrel meets horizontal workings there is a Shaft Station which allows men, materials and services to enter and exit the shaft. From the station tunnels (drifts, galleries or levels) extend towards the ore body, sometimes for many kilometers. The lowest Shaft Station is most often the point where rock leaves the mine levels and is transferred to the shaft, if so a Loading Pocket is excavated on one side of the shaft at this location to allow transfer facilities to be built. Beneath the lowest Shaft Station the shaft continues on for some distance, this area is referred to as the Shaft Bottom. A tunnel called a Ramp typically connects the bottom of the shaft with the rest of the mine, this Ramp often contains the mine's water handling facility, called the Sump, as water will naturally flow to the lowest point in the mine.

SHAFT LINING

The shaft lining performs several functions; it is first and foremost a safety feature preventing loose or unstable rock from falling into the shaft, then a place for shaft sets to bolt into and lastly a smooth surface to minimise resistance to airflow for ventilation. In North and South America, smaller shafts are designed to be rectangular with timber supports. Larger shafts are round and are concrete lined. Final choice of shaft lining is dependent on the geology of the rock which the shaft passes through, some shafts have several liners sections as required Where shafts are sunk in very competent rock there may be no requirement for lining at all, or just the installation of welded mesh and rock bolts. The material of choice for shaft lining is mass concrete which is poured behind Shaft Forms in Lifts of 6m as the shaft advances (gets deeper). Shotcrete, fibrecrete, brick, cast iron tubing, precast concrete segments have all been used at one time or another. Additionally, the use of materials like Bitumen and even squash balls have been required by specific circumstances. In extreme circumstances, particularly when sinking through Halite, composite liners consisting of two or more materials may be required.

SHAFT COMPARTMENTS

Fig. Typical mine cage, located in Harzbergbau, Germany

Where the shaft is to be used for hoisting it is frequently split into multiple compartments by Shaft Sets, these may be made of eithertimber or steel.

Vertical members in a shaft set are called Guides, horizontal members are called Buntons. For steel shaft guides, the main two options are hollow structural sections and top hat sections. Top hat sections offer a number of advantages of hollow structural sections including simpler installation, improved corrosion resistance and increased stiffness. Mine conveyances run on the guides in a similar way to how a steel roller coaster runs on its rails, both having wheels which keep them securely in place.

Some shafts do not use guide beams but instead utilize steel wire rope kept in tension by massive weights at shaft bottom called cheese weights (because of their resemblance to a truckle or wheel of cheese) as these are easier to maintain and replace. The largest compartment is typically used for the mine cage, a conveyance used for moving workers and supplies below the surface, which is suspended from the hoist on steel wire rope. It functions in a similar manner to an elevator. Cages may be single-, double-, or rarely triple-deck, always having multiple redundant safety systems in case of unexpected failure.

The second compartment is used for one or more skips, used to hoist ore to the surface. Smaller mining operations use a skip mounted underneath the cage, rather than a separate device, while some large mines have separate shafts for the cage and skips. The third compartment is used for an emergency exit; it may house an auxiliary cage or a system of ladders. An additional compartment houses mine services such as high voltage cables and pipes for transfer ofwater, compressed air or diesel fuel. A second reason to divide the shaft is for ventilation. One or more of the compartments discussed above may be used for air intake, while others may be used for exhaust. Where this is the case a steel wall called a Brattice is installed between the two compartments to separate the air flow.

SLOPE MINING

Slope mining is a method of accessing valuable geological material, such as coal or ore. A sloping access shaft travels downwards towards desired material. Slope mines differ from shaft and drift mines, which access resources by tunneling straight down or horizontally, respectively. In slope mining, the primary access to the mine is on an incline. Mine hoists may still be used to raise and lower loads on the incline if it is steep, but on shallower slopes, conveyor belts, locomotives or trucks may do the work. Drainage and ventilation of slope mines may be done using the primary slope, or it may be done using auxiliary shafts or bore-holes.

RETREAT MINING

Retreat mining is a term used to reference the final phase of an underground mining technique known as room and pillar mining. This involves excavating a room or chamber while leaving behind pillars of material for support. This excavation is carried out in a pattern advancing away from the entrance of a mine. Once a deposit has been exhausted using this method, the pillars that were left behind initially are removed, or "pulled", retreating back towards the mine's entrance. After the pillars are removed, the roof (or back) is allowed to collapse behind the mining area. Pillar removal must occur in a very precise order in order to reduce the risks to workers, due to the high stresses placed on the remaining pillars by the abutment stresses of the caving ground. Retreat mining is a particularly dangerous form of mining: according to the Mine Safety and Health Administration (MSHA), pillar recovery mining has been historically responsible for 25% of American coal mining deaths, even though it represents only 10% of the coal mining industry.

Retreat mining is the term commonly referred to as the mining method employed by the Crandall Canyon Mine and is believed, by some, to be a factor in the August 2007 collapse which trapped six miners. On November 21, 2007, the mine was sealed, entombing the deceased miners. The bureau of land management says retreat mining was being used. Robert E. Murray, CEO of Murray Energy (owner of the mine), says the technique was not being used at Crandall Canyon.

LONGWALL MINING

Fig. Longwall mining

Longwall mining is a form of underground coal mining where a long wall of coal is mined in a single slice (typically 0.6–1.0 m thick). The longwall *panel* (the block of coal that is being mined) is typically 3–4 km long and 250–400 m wide.

HISTORY

The basic idea of longwall mining was developed in England in the late

17th century. Miners would undercut the coal along the width of the coal face, removing coal as it fell, and using wooden props to control the fall of the roof behind the face. this was known as the *Shropshire method* of mining. While the technology has changed considerably, the basic idea remains the same, to remove essentially all of the coal from a broad coal face and allow the roof and overlying rock to collapse into the void behind, while maintaining a safe working space along the face for the miners.

Starting around 1900, mechanization was applied to this method. By 1940, some referred to longwall mining as "the conveyor method" of mining, after the most prominent piece of machinery involved. Unlike earlier longwall mining, the use of aconveyor belt parallel to the coal face forced the face to be developed along a straight line.

The only other machinery used was an electric cutter to undercut the coal face and electric drills for blasting to drop the face. Once dropped, manual labor was used to load coal onto the conveyor parallel to the face and to place wooden roof props to control the fall of the roof. Such low-technology longwall mines continued in operation into the 1970s.

The best known example was the New Gladstone Mine near Centerville, Iowa "one of the last advancing longwall mines in the United States". This longwall mine did not even use a conveyor belt, but relied on ponies to haul coal tubs from the face to the slope where a hoist hauled the tubs to the surface. Longwall mining has been extensively used as the final stage in mining old room and pillar mines. In this context, longwall mining can be classified as a form of retreat mining.

LAYOUT

Gate roads are driven to the back of each panel before longwall mining begins. The gate road along one side of the block is called the *maingate* or *headgate*; the road on the other side is called the *tailgate*. Where the thickness of the coal allows, these gate roads have been previously developed by continuous miner units, as the longwall itself is not capable of the initial development. The layout of Longwall could be either 'advancing' type or of 'retreat' type.

In the advancing type, the gate roads are formed as the coal face advances.In thinner seams the advancing longwall mining method may be used. In the retreat type, the panel is formed by driving maingate, tail gate and a face connecting the both. Only the maingate road is formed in advance of the face.

The tailgate road is formed behind the coal face by removing the stone above coal height to form a roadway that is high enough to travel in. The end of the block that includes the longwall equipment is called the face. The other end of the block is usually one of the main travel roads of the mine. The cavity behind the longwall is called the *goaf*, *goff* or *gob*.

Ventilation

Typically, intake (fresh) air travels up the main gate, across the face, and then down the tail gate, known as 'U' type ventilation. Once past the face the air is no longer fresh air, but return air carrying away coal dust and mine gases such as methane, carbon dioxide, depending on thegeology of the coal.

Return air is extracted by ventilation fans mounted on the surface. Other ventilation methods can be used where intake air also passes the main gate and into a bleeder or back return road reducing gas emissions from the goaf onto the face, or intake air travels up the tail gate and across the face in the same direction as the face chain in a homotropal system.

Typically to avoid coal in the goaf area spontaneously combusting, gases may be allowed to build up behind seals so as to exclude oxygen from the sealed goaf area. Where a goaf may contain a explosive mixture of methane and oxygen, nitrogen injection/inertisation may be used to exclude oxygen or push the explosive mixture deep into the goaf where there are no probable ignition sources. Seals are required to be monitored each shift by a certified mine supervisor for damage and leaks of harmful gases.

Equipment

Fig. Hydraulic chocks

A number of hydraulic jacks, called *powered roof supports, chocks* or *shields,* which are typically 1.75 m wide and placed in a long line, side by side for up to 400 m in length in order to support the roof of the coalface. An individual chock can weigh 30–40 tonnes, extend to a maximum cutting height of up to 6 m and have yield rating of 1000–1250 tonnes each, and hydraulically advance itself 1 m at a time.

Fig. Hydraulic chocks, conveyor and shearer

The coal is cut from the coalface by a machine called the *shearer* (*power loader*). This machine can weigh 75–120 tonnes typically and comprises a main body, housing the electrical functions, the tractive motive units to move the shearer along the coalface and pumping units (to power both hydraulic and water functions).

At either end of the main body are fitted the ranging arms which can be ranged vertically up down by means of hydraulic rams, and onto which are mounted the shearer cutting drums which are fitted 40–60 cutting picks. Within the ranging arms are housed very powerful electric motors (typically up to 850 kW) which transfer their power through a series of lay gears within the body the arms to the drum mounting locations at the extreme ends of the ranging arms where the cutting drums are. The cutting drums are rotated at a speed of 20–50 revs/min to cut the mineral from coal seam.

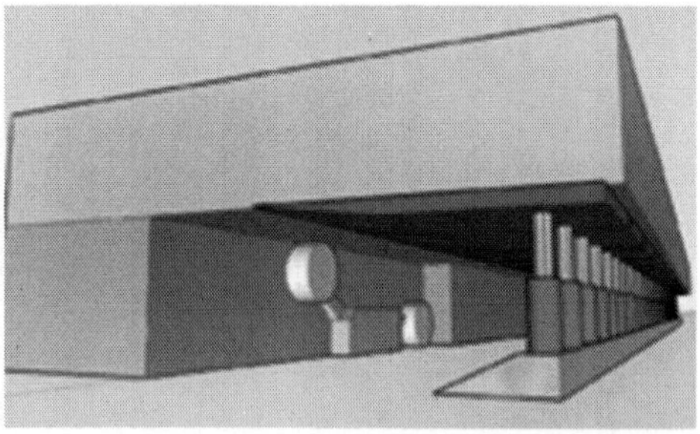

Fig. Chocks providing support to allow shearer to work

The shearer is carried along the length of the face on the *armoured face conveyor* (AFC); using a chain-less haulage system, which resembles a ruggedised rack and pinion system especially developed for mining. Before chainless haulage systems, a heavy duty chain was run the length of the coal face for the shearer to pull itself along. The shearer moves at a speed of 10–30 m/min depending on cutting conditions. The AFC is placed in front of the powered roof supports, and the shearing action of the rotating drums cutting into the coal seam disintegrates the coal, this being loaded onto the AFC. The coal is removed from the coal face by a scraper chain conveyor to the main gate. Here it is loaded onto a network of conveyor belts for transport to the surface. At the main gate the coal is usually reduced in size in a crusher, and loaded onto the first conveyor belt by the *beam stage loader* (BSL).

As the shearer removes the coal, the AFC is snaked over behind the shearer and the powered roof supports move forward into the newly created cavity. As mining progresses and the entire longwall progresses through the seam, the goaf increases. This goaf collapses under the weight of the overlying strata. The strata approximately 2.5 times the thickness of the coal seam removed collapses and the beds above settle onto the collapsed goaf. This collapsing can lower surface height, causing problems like changing the course of rivers and severely damage building foundations.

COMPARISON WITH ROOM AND PILLAR METHOD

Longwall and room and pillar methods of mining can both be used for mining suitable underground coal seams. Longwall has better resource recovery (about 80% compared with about 60% for room and pillar method, fewer roof support consumables are needed, higher volume coal clearance systems, minimal manual handling and safety of the miners is enhanced by the fact that they are always under the hydraulic roof supports when they are extracting coal.

Subsidence

There have been cases of surface subsidence altering the landscape above the mines. At Newstan Colliery in New South Wales, Australia "the surface has dropped by as much as five metres in places" above a multi level mine. In some cases the subsidence causes damage to natural features such as drainage to water courses or man-made structures such as roads and buildings. "Douglas Park Drive was closed for four weeks because longwall panels ... destabilised the road. In 2000, the State Government stopped mining when it came within 600 metres from the twin bridges. A year later there were reports of 40-centimetre gaps appearing in the road, and the bridge had to be jacked sideways to realign it."

A 2005 geotechnical report commissioned by the NSW RTA warns that "subsidence could happen suddenly and occur over many years".

However, there are several mines, which were successfully mined with little to no measurable surface subsidence including mines under lakes, oceans, important water catchments and environmentally sensitive areas. Subsidence is minimised by increasing the block's adjacent chain pillar widths, decreasing extracted block widths and heights, and by giving consideration to the depth of cover as well as competency and thickness of overlying strata.

First mining company in Canada to use long-wall mining technique

A coal mine in Tumbler Ridge, northern British Columbia, Canada operated by HD Mining International, a company majority-owned by Huiyong Holdings Group, a private company from China uses long-wall mining in which "coal is extracted along a wall in large blocks and then carried out on a conveyor belt." HD Mining claims theirs is the first mine in Canada to use this technology.

HYDRAULIC MINING

Fig. A miner using a hydraulic jet to mine for gold in California, from *The Century Magazine* January 1883

Hydraulic mining, or hydraulicking, is a form of mining that uses high-pressure jets of water to dislodge rock material or move sediment. In the placer mining of gold or tin, the resulting water-sediment slurry is directed through sluice boxes to remove the gold. It is also used in mining kaolin and coal. Hydraulic mining originated out of ancient Roman techniques that used water to excavate soft underground deposits. Its modern form, using pressurized water jets produced by a nozzle called a "monitor", came about in the 1850s during the California Gold Rush in the United States. Though successful in extracting gold-rich minerals, the widespread use of the process resulted in extensive environmental damage, such as increased flooding and erosion, and sediment blocking waterways and covering over farm fields. These problems lead to its legal regulation. Hydraulic mining has since been used in various forms around the world.

HISTORY

Precursor: ground sluicing

Hydraulic mining had its precursor in the practice of ground sluicing, a development of which is also known as "hushing", in which surface streams of water were diverted so as to erode gold-bearing gravels. This was originally used in the Roman empire in the first centuries AD and BC, and expanded throughout the empire wherever alluvial deposits occurred The Romans used ground sluicing to remove overburden and the gold-bearing debris in Las Médulas of Spain, and Dolaucothi in Britain. The method was also used in Elizabethan Britain for developing lead, tin and copper mines.

Fig. Panoramic view of Las Médulas

Water was used on a large scale by Roman engineers in the first centuries BC and AD when the Roman empire was expanding rapidly in Europe. Using a process later known as hushing, the Romans stored a large volume of water in a reservoir immediately above the area to be mined; the water was then quickly released.

The resulting wave of water removed overburden and exposed bedrock. Gold veins in the bedrock were then worked using a number of techniques, and water power was used again to remove debris. The remains at Las Medulas and in surrounding areas show badland scenery on a gigantic scale owing to hydraulicking of the rich alluvial gold deposits. Las Medulas is now a UNESCO World Heritage site. The site shows the remains of at least seven large aqueducts of up to 30 miles in length feeding large supplies of water into the site.

The gold-mining operations were described in vivid terms by Pliny the Elder in his Naturalis Historia published in the first century AD. Pliny was a procurator in Hispania Terraconensis in the 70's and must have witnessed for himself the operations. The use of hushing has been confirmed by field survey and archaeology at Dolaucothi in South Wales, the only known Roman gold mine in Britain.

California Gold Rush

The modern form of hydraulic mining, using jets of water directed under very high pressure through hoses and nozzles at gold-bearing upland paleogravels, was first used by Edward Matteson near Nevada City, California in 1853 during the California Gold Rush. Matteson used canvas hose which was later replaced with crinoline hose by the 1860s.

In California, hydraulic mining often brought water from higher locations for long distances to holding ponds several hundred feet above the area to be mined. California hydraulic mining exploited gravel deposits, making it a form of placer mining.

Fig. Gold miners excavate an eroded bluff with jets of water at a placer mine in Dutch Flat, California sometime between 1857 and 1870.

Early placer miners in California discovered that the more gravel they could process, the more gold they were likely to find. Instead of working with pans, sluice boxes, long toms, and rockers, miners collaborated to find ways to process larger quantities of gravel more rapidly.

Hydraulic mining became the largest-scale, and most devastating, form of placer mining. Water was redirected into an ever-narrowing channel, through a large canvas hose, and out through a giant iron nozzle, called a "monitor." The extremely high pressure stream was used to wash entire hillsides through enormous sluices.

By the early 1860s, while hydraulic mining was at its height, small-scale placer mining had largely exhausted the rich surface placers, and the mining industry turned to hard rock (called quartz mining in California) or hydraulic mining, which required larger organizations and much more capital. By the mid-1880s, it is estimated that 11 million ounces of gold (worth approximately US$7.5 billion at mid-2006 prices) had been recovered by hydraulic mining in the California Gold Rush.

Environmental Consequences

While generating millions of dollars in tax revenues for the state and supporting a large population of miners in the mountains, hydraulic mining had a devastating effect on riparian natural environment and agricultural systems in California.

Millions of tons of earth and water were delivered to mountain streams that fed rivers flowing into the Sacramento Valley. Once the rivers reached the relatively flat valley, the water slowed, the rivers widened, and the sediment was deposited in the floodplains and river beds causing them to rise, shift to new channels, and overflow their banks, causing major flooding, especially during the spring melt.

Fig. A man leans over a wooden sluice. Rocks line the outside of the wood boards that create the sluice.

Cities and towns in the Sacramento Valley experienced an increasing number of devastating floods, while the rising riverbeds made navigation on the rivers increasingly difficult.

Perhaps no other city experienced the boon and the bane of gold mining as much as Marysville. Situated at the confluence of the Yuba and Feather rivers, Marysville was the final "jumping off" point for miners heading to the northern foothills to seek their fortune. Steamboats from San Francisco, carrying miners and supplies, navigated up the Sacramento River, then the Feather River to Marysville where they would unload their passengers and cargo.

Marysville eventually constructed a complex levee system to protect the city from floods and sediment. Hydraulic mining greatly exacerbated the problem of flooding in Marysville and shoaled the waters of the Feather River so severely that few steamboats could navigate from Sacramento to the Marysville docks. The sediment left by such efforts were reprocessed by mining dredges at the Yuba Goldfields, located near Marysville.

Fig. The Malakoff Diggins, California, showing the effects of hydraulic mining on a hillside over a century later. Much of the effects of the mining was beyond the hills themselves, on the areas downstream of the water and sediment flow they produced.

The spectacular eroded landscape left at the site of hydraulic mining can be viewed at Malakoff Diggins State Historic Park inNevada County, California. A similar landscape can be seen at Las Médulas in northern Spain, where Roman engineers ground sluiced the rich gold alluvial deposits of the river Sil. Pliny the Elder mentions in his Naturalis Historia that Spain had encroached on the sea and local lakes as a result of ground sluicing operations.

Legal consequences

Vast areas of farmland in the Sacramento Valley were deeply buried by the mining sediment. Frequently devastated by flood waters, farmers demanded an end to hydraulic mining. In the most renowned legal fight of farmers against miners, the farmers sued the hydraulic mining operations and the landmark case of *Edwards Woodruff v.*

*North Bloomfield Mining and Gravel Company*made its way to the United States District Court in San Francisco where Judge Lorenzo Sawyer decided in favor of the farmers in 1884, declaring that hydraulic mining was "a public and private nuisance" and enjoining its operation in areas tributary to navigable streams and rivers.

Hydraulic mining was recommenced after 1893 when the United States Congress passed the Camminetti Act which allowed such mining if sediment detention structures were constructed. This led to a number of operations above brush dams and log crib dams. Most of the water-delivery infrastructure had been destroyed by an 1891 flood, so this later stage of mining was carried on at a much smaller scale in California.

Beyond California

Fig. The Oriental Claims near Omeo, Australia were mined between the 1850s and 1900s; hydraulic sluicing left man-made cliffs up to 30 metres (98 ft) high such as seen here throughout the area

Although often associated with California due to its adoption and widespread use there, the technology was exported widely, toOregon (Jacksonville in 1856), Colorado (Clear Creek, Central City and Breckenridge in 1860), Montana (Bannack in 1865),Arizona (Lynx Creek in 1868), Idaho (Idaho City in 1863), South Dakota (Deadwood in 1876), Alaska, British Columbia (Canada), and overseas.

It was used extensively in Dahlonega, Georgia and continues to be used in developing nations, often with devastating environmental consequences. The devastation caused by this method of mining caused Edwin Carter, the "Log Cabin Naturalist," to switch from mining to collecting wildlife specimens from 1875-1900 in Breckenridge, Colorado, USA.

Hydraulic mining was also used during the Australian gold rushes where it was called hydraulic sluicing. One notable location was at the Oriental Claims near Omeo in Victoria where it was used between the 1850s and early 1900s, with abundant evidence of the damage still being visible today. Hydraulic mining was used extensively in the Central Otago Gold Rush that took place in the 1860s in the South Island of New Zealand, where it was also known as *sluicing*.

Starting in the 1870s, hydraulic mining became a mainstay of alluvial tin mining on the Malay Peninsula. Hydraulicking was formerly used in Polk County, Florida to mine phosphate rock.

CONTEMPORARY USAGE

Fig. Modern hydraulic mining monitor in use

In addition to its use in true mining, hydraulic mining can be used as an excavation technique, principally to demolish hills. For example, the Denny Regrade in Seattle was largely accomplished by hydraulic mining. Hydraulic mining is the principal way that kaolinite clay is mined in Cornwall and Devon, in South-West England.

On the South African Rand gold fields, a gold surface tailings re-treatment facility called East Rand Gold and Uranium Company (ERGO) has been in operation since 1977 using hydraulic monitors to create slurry from older (and consequently richer) tailings sites and pumping these long distances to a concentration plant. The facilities process nearly 2.0 million tons of tailings each month at a processing cost of below USD 3.00/t (2013). Gold is recovered at a rate of only 0.20g/t, but the low yield is compensated for by the extremely low cost of processing - no risky or expensive mining or milling is required for recovery. The resulting, new, slimes are pumped further away from the built up areas permitting the economic development of land close to commercially valuable areas and previously covered by the tailings. The historic yellow-coloured mine dumps around Johannesburg are now almost a rarity seen only in older photographs. Uranium and pyrites (for sulphuric acid production) are also available for recovery from the process stream as co-products under suitable economic conditions.

Underground hydraulic mining

High-pressure water jets have also been used in the underground mining of coal, to break up the coal seam and wash the resulting coal slurry toward a collection point. The high-pressure water nozzle is referred to as the 'hydro monitor'.

COPPER EXTRACTION TECHNIQUES

Fig. Fog/ Chalcopyrite

Copper extraction techniques refers to the methods for obtaining copper from its ores. The conversion of copper consists of a series of chemical, physical, and electrochemical processes. Methods have evolved and vary with country depending on the ore source, local environmental regulations, and other factors. As in all mining operations, the ore must usually be beneficiated (concentrated). To do this, the ore is crushed. Then it must beroasted to convert sulfides to oxides, which are smelted to produce matte. Finally, it undergoes various refining processes, the final one being electrolysis. For economic and environmental reasons, many of the byproducts of extraction are reclaimed. Sulfur dioxidegas, for example, is captured and turned into sulfuric acid — which is then used in the extraction process.

HISTORY

The earliest evidence of cold-hammering of native copper comes from the excavation at Çayönü Tepesi in eastern Anatolia. The radiocarbon date is 7250 ± 250 BCE. Among the various items considered to be votive or amulets there was one that looked like a fishhook and one like an awl. An archaeological site in southeastern Europe (Serbia) contains the oldest securely dated evidence of copper making at high temperature, from 7,000 years ago. The find in June 2010 extends the known record of copper smelting by about 500 years, and suggests that copper smelting may have been invented in separate parts of Asia and Europe at that time rather than spreading from a single source. Copper smelting technology gave rise to the Copper Age and then the Bronze Age.

Concentration

Most copper ores contain only a small percentage of copper metal bound up within valuable ore minerals, with the remainder of the ore being unwanted rock or gangueminerals, typically silicate minerals or oxide minerals for which there is often no value. The average grade of copper ores in the 21st century is below 0.6% copper, with a proportion of economic ore minerals (including copper) being less than 2% of the total volume of the ore rock. A key objective in the metallurgical treatment of any ore is the separation of ore minerals from gangue minerals within the rock. The first stage of any process within a metallurgical treatment circuit is accurate grinding or *comminution*, where the rock is crushed to produce small particles (<100 ìm) consisting of individual mineral phases. These particles are then separated to remove gangue, thereafter followed by a process of physical liberation of the ore minerals from the rock. The process of liberation of copper ores depends upon whether they are oxide or sulfide ores. Subsequent steps depend on the nature of the ore containing the copper. For oxide ores, a hydrometallurgical liberation process is normally undertaken, which uses the soluble nature of the ore minerals to the advantage of the metallurgical treatment plant. For sulfide ores, both secondary (supergene) and primary (unweathered), froth flotation is used to physically separate ore from gangue. For special native copper bearing ore bodies or sections of ore bodies rich in supergen native copper, this mineral can be recovered by a simple gravity circuit.

Froth flotation

Fig. Froth flotation cells to concentrate copper and nickel sulfide minerals, Falconbridge, Ontario.

The modern froth flotation process was independently invented the early 1900s in Australia by C.V Potter and around the same time by G. D. Delprat.

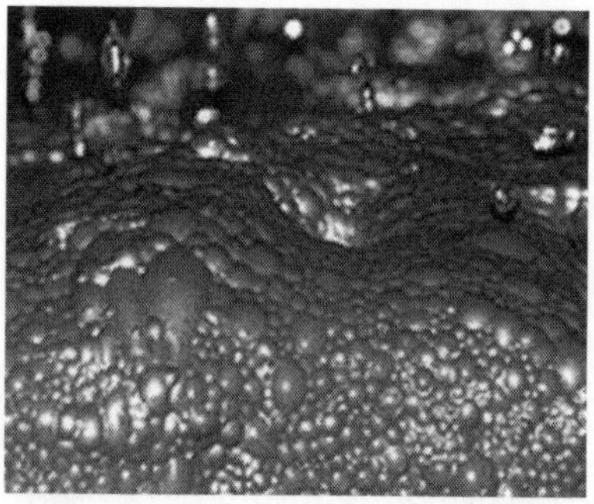

Fig. Copper sulfide loaded air bubbles on aJameson Cell at the flotation plant of theProminent Hill mine in South Australia

All primary sulfide ores of copper sulfides, and most concentrates of secondary copper sulfides (being chalcocite), are subjected tosmelting. Some vat leach or pressure leach processes exist to solubilise chalcocite concentrates and produce copper cathode from the resulting leachate solution, but this is a minor part of the market. Carbonate concentrates are a relatively minor product produced from copper cementation plants, typically as the end-stage of a heap-leach operation.

Such carbonate concentrates can be treated by a solvent extraction and electrowinning (SX-EW) plant or smelted. The copper ore is crushed and ground to a size such that an acceptably high degree of liberation has occurred between the copper sulfide ore minerals and the gangue minerals. The ore is then wet, suspended in a slurry, and mixed with xanthates or other reagents, which render the sulfide particles hydrophobic. Typical reagents include potassium ethylxanthate and sodium ethylxanthate, but dithiophosphates and dithiocarbamates are also used.

The treated ore is introduced to a water-filled aeration tank containing surfactant such as methylisobutyl carbinol (MIBC). Air is constantly forced through the slurry and the air bubbles attach to the hydrophobic copper sulfide particles, which are conducted to the surface, where they form a froth and are skimmed off. These skimmings are generally subjected to a cleaner-scavenger cell to remove excess silicates and to remove other sulfide minerals that can deleteriously impact the concentrate quality (typically, galena), and the final concentrate sent for smelting. The rock which has not floated off in the flotation cell is either discarded astailings or further processed to extract other metals such as lead (from galena) and zinc (from sphalerite), should they exist. To improve the process efficiency, lime is used to raise the pH of the water bath, causing the collector to ionize more and to preferentially bond to chalcopyrite ($CuFeS_2$) and avoid the pyrite (FeS_2). Iron exists in both primary zone minerals.

Copper ores containing chalcopyrite can be concentrated to produce a concentrate with between 20% and 30% copper-in-concentrate (usually 27–29% copper); the remainder of the concentrate is iron and sulfur in the chalcopyrite, and unwanted impurities such as silicate gangue minerals or other sulfide minerals, typically minor amounts of pyrite, sphalerite or galena. Chalcocite concentrates typically grade between 37% and 40% copper-in-concentrate, as chalcocite has no iron within the mineral.

HYDROMETALLURGICAL EXTRACTION

Sulfide Ores

Secondary sulfides – those formed by supergene secondary enrichment – are resistant (*refractory*) to sulfuric leaching. These ores are a mixture of copper carbonate, sulfate, phosphate, and oxide minerals and secondary sulfide minerals, dominantly chalcocite but other minerals such as digenite can be important in some deposits. Supergene ores rich in sulfides may be concentrated using froth flotation. A typical concentrate of chalcocite can grade between 37% and 40% copper in sulfide, making them relatively cheap to smelt compared to chalcopyrite concentrates.

Some supergene sulfide deposits can be leached using a bacterial oxidation heap leach process to oxidize the sulfides to sulfuric acid, which also allows for simultaneous leaching with sulfuric acid to produce a copper sulfate solution. As with oxide ores, solvent extraction and electrowinning technologies are used to recover the copper from thepregnant leach solution. Supergene sulfide ores rich in native copper minerals are refractory to treatment with sulfuric acid leaching on all practicable time scales, and the dense metal particles do not react with froth flotation media. Typically, if native copper is a minor part of a supergene profile it will not be recovered and will report to the tailings. When rich enough, native copper ore bodies may be treated to recover the contained copper via a gravity separation circuit where the density of the metal is used to liberate it from the lighter silicate minerals. Often, the nature of the gangue is important, as clay-rich native copper ores prove difficult to liberate.

Oxide ores

Oxidised copper ore bodies may be treated via several processes, with hydrometallurgical processes used to treat oxide ores dominated by copper carbonate minerals such as azurite and malachite, and other soluble minerals such as silicates like chrysocolla, or sulfates such as atacamite and so on. Such oxide ores are usually leached by sulfuric acid, usually in a heap leaching or dump leaching process to liberate the copper minerals into a solution of sulfuric acid laden with copper sulfate in solution. The copper sulfate solution (the pregnant leach solution) is then stripped of copper via a solvent extraction and electrowinning (SX-EW) plant, with the barred (denuded) sulfuric acid recycled back on to the heaps.

Alternatively, the copper can be precipitated out of the pregnant solution by contacting it with scrap iron; a process called cementation. Cement copper is normally less pure than SX-EW copper. Commonly sulfuric acid is used as a leachant for copper oxide, although it is possible to use water, particularly for ores rich in ultra-soluble sulfate minerals.

In general, froth flotation is not used to concentrate copper oxide ores, as oxide minerals are not responsive to the froth flotation chemicals or process (*i.e.;* they do not bind to the kerosene-based chemicals). Copper oxide ores have occasionally been treated via froth flotation via sulfidation of the oxide minerals with certain chemicals which react with the oxide mineral particles to produce a thin rime of sulfide (usually chalcocite), which can then be activated by the froth flotation plant.

SULFIDE SMELTING

Until the latter half of the 20th century, smelting sulfide ores was almost the sole means of producing copper metal from mined ores (*primary* copper production). Davenport, et al, noted in 2002 that even then 80% of global primary copper production was from copper–iron–sulfur minerals and that the vast majority of these were treated by smelting.

Copper was initially recovered from sulfide ores by directly smelting the ore in a furnace. The smelters were initially located near the mines to minimize the cost of transport. This avoided the prohibitive costs of transporting the waste minerals and the sulfur and iron present in the copper-containing minerals. However, as the concentration of copper in the ore bodies decreased, the energy costs of smelting the whole ore also became prohibitive, and it became necessary to concentrate the ores first.

Initial concentration techniques included hand-sorting and gravity concentration. They resulted in high losses of copper. Consequently, the development of the froth flotation process was a major step forward in mineral processing. It made possible the development of the giant Bingham Canyon mine in Utah. In the twentieth century, most ores were concentrated before smelting. Smelting was initially undertaken using sinter plants andblast furnaces, or with roasters and reverberatory furnaces. Roasting and reverberatory furnace smelting dominated primary copper production until the 1960s.

Roasting

The roasting process is generally undertaken in combination with reverberatory furnaces. In the roaster, the copper concentrate is partially oxidised to produce "calcine" and sulfur dioxide gas. The stoichiometry of the reaction which occurs is:

$$2\ CuFeS_2 + 3\ O_2 \rightarrow 2\ FeO + 2\ CuS + 2\ SO_2$$

Roasting generally leaves more sulfur in the calcined product (15% in the case of the roaster at Mount Isa Mines) than a sinter plant leaves in the sintered product (about 7% in the case of the Electrolytic Refining and Smelting smelter). As of 2005, roasting is no longer common in copper concentrate treatment, because its combination with reverberatory furnaces is not energy efficient and the SO_2 concentration in the roaster offgas is too dilute for cost-effective capture. Direct smelting is now favored, *e.g.* using the following smelting technologies: ISASMELT, flash smelting, Noranda, Top-entry Submerged Lance, Mitsubishi or El Teniente furnaces.

Smelting

The initial melting of the material to be smelted is usually referred to as the *smelting* or *matte smelting* stage. It can be undertaken in a variety of furnaces, including the largely obsolete blast furnaces, reverberatory furnaces, flash furnaces, Isasmelt furnaces, etc. The product of this smelting stage is a mixture of copper, iron and sulfur that is enriched in copper, and which is called *matte* or *copper matte*. The term *matte grade* is normally used to refer to the copper content of the matte.

The purpose of the matte smelting stage is to eliminate as much of the unwanted iron, sulfur and *gangue* minerals (such as silica, magnesia, alumina and limestone) as possible, while minimizing the loss of copper. This is achieved by reacting iron sulfides with oxygen (in air or oxygen enriched air) to produce iron oxides (mainly as FeO, but with some magnetite (Fe_3O_4)) and sulfur dioxide. Copper sulfide and iron oxide can mix, but when sufficient silica is added, a separate slag layer is formed. Adding silica also reduces the melting point (or, more properly, the liquidus temperature) of the slag, meaning that the smelting process can be operated at a lower temperature.

The slag forming reaction is:

$$FeO + SiO_2 \rightarrow FeO.SiO_2$$

Slag is less dense than matte, so it forms a layer that floats on top of the matte.

Copper can be lost from the matte in three ways: as cuprous oxide (Cu_2O) dissolved in the slag, as sulfide copper dissolved in the slag or as tiny droplets (or *prills*) of matte suspended in the slag. The amount of copper lost as oxide copper increases as the oxygen potential of the slag increases. The oxygen potential generally increases as the copper content of the matte is increased. Thus the loss of copper as oxide increases as the copper content of the matte increases.

On the other hand, the solubility of sulfidic copper in slag decreases as the copper content of the matte increases beyond about 40%. Nagamori calculated that more than half the copper dissolved in slags from mattes containing less than 50% copper is sulfidic copper. Above this figure, oxidic copper begins to dominate. The loss of copper as prills suspended in the slag depends on the size of the prills, the viscosity of the slag and the settling time available. Rosenqvist suggested that about half the copper losses to slag were due to suspended prills. The mass of slag generated in the smelting stage depends on the iron content of the material fed into the smelting furnace and the target matte grade. The greater the iron content of the feed, the more iron that will need to be rejected to the slag for a given matte grade. Similarly, increasing the target matte grade requires the rejection of more iron and an increase in the slag volume.

Thus, the two factors that most affect the loss of copper to slag in the smelting stage are:

- Matte grade
- Mass of slag.

This means that there is a practical limit on how high the matte grade can be if the loss of copper to slag can be minimized. Therefore, further stages of processing (converting and fire refining) are required. The following subsections briefly describe some of the processes used in matte smelting.

Reverberatory furnace smelting

Reverberatory furnaces are long furnaces can treat wet, dry or roasted concentrate. Most of the reverberatory furnaces used in the latter years treated roasted concentrate because putting dry feed materials into the reverberatory furnace is more energy efficient, and because the elimination of some of the sulfur in the roaster results in higher matte grades. The reverberatory furnace feed is added to the furnace through feed holes along the sides of the furnace. Additional silica is normally added to help form the slag. The furnace is fired with burners using pulverized coal, fuel oil or natural gas and the solid charge is melted.

Reverberatory furnaces can additionally be fed with molten slag from the later converting stage to recover the contained copper and other materials with a high copper content. Because the reverberatory furnace bath is quiescent, very little oxidation of the feed occurs. It is essentially a melting process. Consequently, wet-charged reverberatory furnaces have less copper in their matte product than calcine-charged furnaces, and they also have lower copper losses to slag. Gill quotes a copper in slag value of 0.23% for a wet-charged reverberatory furnace vs 0.37% for a calcine-charged furnace.

In the case of calcine-charged furnaces, a significant portion of the sulfur has been eliminated during the roasting stage, and the calcine consists of a mixture of copper and iron oxides and sulfides. The reverberatory furnace acts to allow these species to approach chemical equilibrium at the furnace operating temperature (about 1600°C at the burner end of the furnace and about 1200°C at the flue end; the matte is about 1100°C and the slag is about 1195°C). In this equilibration process, oxygen associated with copper compounds exchanges with sulfur associated with iron compounds, increasing the iron oxide content of the furnace, and the iron oxides interact with silica and other oxide materials to form the slag.

The main equilibration reaction is:

$$Cu_2O + FeS = Cu_2S + FeO$$

The slag and the matte form distinct layers that can be removed from the furnace as separate streams. The slag layer is periodically allowed to flow through a hole in the wall of the furnace above the height of the matte layer. The matte is removed by draining it through a hole into ladles for it to be carried by crane to the converters. This draining process is knows as *tapping* the furnace. The matte taphole is normally cut a hole through a water-cooled copper block that prevents erosion of the refractory bricks lining the furnace. When the removal of the matte or slag is complete, the hole is normally plugged with clay, which is removed when the furnace is ready to be tapped again.

Reverberatory furnaces were often used to treat molten converter slag to recover contained copper. This would be poured into the furnaces from ladles carried by cranes. However, the converter slag is high in magnetite and some of this magnetite would precipitate from the converter slag (due to its higher melting point), forming an accretion on the hearth of the reverberatory furnace and necessitating shut downs of the furnace to remove the accretion. This accretion formation limits the quantity of converter slag that can be treated in a reverberatory furnace.

While reverberatory furnaces had very low copper losses to slag, they are not very energy-efficient and the low concentrations of sulfur dioxide in their offgases made its capture uneconomic. Consequently, smelter operators devoted a lot of money in the 1970s and 1980s to developing new, more efficient copper smelting processes. In addition, flash smelting technologies had been developed in earlier years and began to replace reverberatory furnaces. By 2002, 20 of the 30 reverberatory furnaces still operating in 1994 had been shut down.

Flash furnace smelting

In flash smelting, the concentrate is dispersed in an air or oxygen stream and the smelting reactions are largely completed while the mineral particles are still in flight. The reacted particles then settle in a bath at the bottom of the furnace, where they behave as does calcine in a reverberatory furnace. A slag layer forms on top of the matte layer, and they can separated be tapped from the furnace.

Converting

Fig. Tough-pitch copper (ca. 98 % pure), containing antimony and nickel

The matte, which is produced in the smelter, contains around 30–70% copper (depending on the process used and the operating philosophy of the smelter), primarily as copper sulfide, as well as iron sulfide. The sulfur is removed at high temperature as sulfur dioxide by blowing air through molten matte:

$$2\ CuS + 3\ O_2 \rightarrow 2\ CuO + 2\ SO_2$$

$$CuS + O_2 \rightarrow Cu + SO_2$$

In a parallel reaction the iron sulfide is converted to slag:

$$2\ FeS + 3\ O_2 \rightarrow 2\ FeO + 2\ SO_2$$

$$2\ FeO + SiO_2 \rightarrow Fe_2SiO_4$$

The purity of this product is 98%, it is known as *blister* because of the broken surface created by the escape of sulfur dioxide gas as blister copper *pigs* or ingots are cooled. By-products generated in the process are sulfur dioxide and slag. The sulfur dioxide is captured for use in earlier leaching processes.

Fire refining

The blister copper is put into an anode furnace, a furnace that refines the blister copper to anode-grade copper in two stages by removing most of the remaining sulfur and iron, and then removing oxygen introduced during the first stage. This second stage, often referred to as *poling* is done by blowing natural gas, or some other reducing agent, through the molten copper oxide. When this flame burns green, indicating the copper oxidation spectrum, the oxygen has mostly been burned off. This creates copper at about 99% pure. The anodes produced from this are fed to the electrorefinery.

Electrorefining

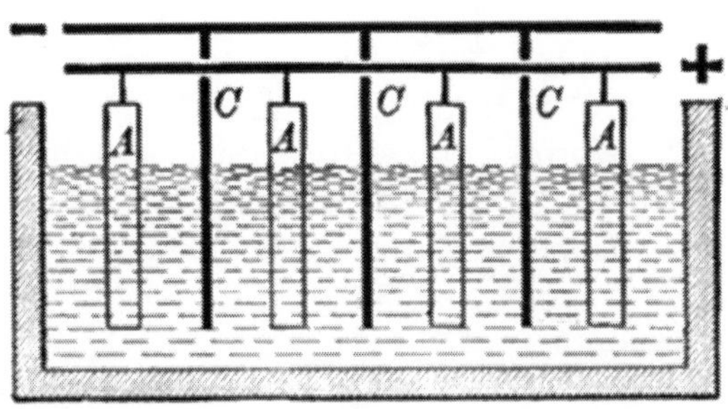

Fig. Apparatus for electrolytic refining of copper

The copper is refined by electrolysis. The anodes cast from processed blister copper are placed into an aqueoussolution of 3–4% copper sulfate and 10–16% sulfuric acid. Cathodes are thin rolled sheets of highly pure copper or, more commonly these days, reusable stainless steel starting sheets (as in the IsaKidd process). A potential of only 0.2–0.4 volts is required for the process to commence. At the anode, copper and less noble metals dissolve. More noble metals such as silver and gold as well as selenium and tellurium settle to the bottom of the cell as anode slime, which forms a saleable byproduct. Copper(II) ions migrate through the electrolyte to the cathode. At the cathode, copper metal plates out but less noble constituents such as arsenic and zinc remain in solution. The reactions are:

$$\text{At the anode: } Cu_{(s)} \rightarrow Cu_{(aq)} + 2e$$
$$\text{At the cathode: } Cu^{2+}_{(aq)} + 2e^{-} \rightarrow Cu_{(s)}$$

CONCENTRATE AND COPPER MARKETING

Copper concentrates produced by mines are sold to smelters and refiners who treat the ore and refine the copper and charge for this service via treatment charges (TCs) and refining charges (RCs). The TCs are charged in US$ per tonne of concentrate treated and RCs are charged in cents per pound treated, denominated in US dollars, with benchmark prices set annually by major Japanese smelters. The customer in this case can be a smelter, who on-sells blister copper ingots to a refiner, or a smelter-refiner which is vertically integrated. The typical contract for a miner is denominated against the London Metal Exchange price, minus the TC-RCs and any applicable penalties or credits. Penalties may be assessed against copper concentrates according to the level of deleterious elements such as arsenic, bismuth, lead or tungsten. Because a large portion of copper sulfide ore bodies contain silver or gold in appreciable amounts, a credit can be paid to the miner for these metals if their concentration *within the concentrate* is above a certain amount. Usually the refiner or smelter charges the miner a fee based on the concentration; a typical contract will say a credit is due for every ounce of the metal in concentrate above a certain concentration; below that, if it is recovered, the smelter will keep the metal and sell it to defray costs.

Copper concentrate is traded either via spot contracts or under long term contracts as an intermediate product in its own right. Often the smelter sells the copper metal itself on behalf of the miner. The miner is paid the price at the time that the smelter-refiner makes the sale, not at the price on the date of delivery of the concentrate. Under a Quotational Pricing system, the price is agreed to be at a fixed date in the future, typically 90 days from time of delivery to the smelter.

A-grade copper cathode is of 99.99% copper in sheets that are 1 cm thick, and approximately 1 meter square weighing approximately 200 pounds. It is a true commodity, deliverable to and tradeable upon the metal exchanges in New York (COMEX), London (London Metals Exchange) and Shanghai (Shanghai Futures Exchange). Often copper cathode is traded upon the exchanges indirectly via warrants, options, or swap contracts such that the majority of copper is traded upon the LME/COMEX/SFE but delivery is achieved indirectly and at remove from the physical warehouses themselves. The chemical specification for electrolytic grade copper is ASTM B 115-00 (a standard that specifies the purity and maximum electrical resistivity of the product).

4

Mining Equipment

Mining industry equipment has particular modifications that are required for the individual shaft environments of each underground project. Miners and manufacturers of mining equipment agree that safety is equal to output capability, but neither should have to be sacrificed. Mining equipment maintenance should be conducted by professionals familiar with the equipment and common problems with it, as well as solution based problem solving skills to handle modification needs for this type of machinery. Mining equipment maintenance also requires a knowledge of the entire mining process, so that materials are processed optimally at all stages, from excavation to product output.

THE MODERN ART OF LUBRICATING MINING EQUIPMENT

Surface mining is a very old practice in North America. For example, mining for copper at the surface of the Michigan Upper Peninsula appears to have started as early as 5000 B.C. The ore produced through the hard labor of thousands of miners went to the manufacture of hand tools and jewelry. It is estimated by scientists and engineers that given the tools available at the time, it would have taken 10,000 people approximately 1,000 years to excavate the 1.5 billion pounds of ore from this area of Michigan.

In contrast, it would take approximately 38.5 hours for the fabled Big Muskie dragline to excavate the same amount of material. Perhaps if the Chippewa had the Big Muskie on hand they would not have given up the 30,000 square miles of the Upper Peninsula, with its high-quality copper from Lake Superior.

Talk of surface mining conjures visions of panning in a river during the California gold rush period in 1849. Since that time, North Americans have seen tremendous changes in many areas of life, not the least of which is the surface mining industry. Mining equipment has evolved into some of the most fascinating functional mechanical structures known to man.

Fig. North American's Oldest Dragline

As a case in point, take a look at this Bucyrus Class 24, currently being restored at Reynolds Alberta Museum, located in Wetaskiwin, Alberta, Canada. This steam-powered dragline, erected in 1917, utilized a 100-foot boom with a four-yard bucket. By comparison, a modern standard size pick-up truck offers about 2.5 cubic yards of space. To move it from one location to another, timbers were placed under the tub and the operator would use the bucket to pull the machine over the timbers to the desired location.

Fig. Bucyrus International, Inc. 2570WS

From this primitive beginning, mining tools have evolved into massive machines. Rock drills, front-end loaders, electric cable and hydraulic shovels - some boasting bucket sizes in excess of 50 yards - and mammoth trucks and draglines, such as the Bucyrus International, Inc. 2570WS, have introduced unprecedented levels of productivity for equipment owners and unprecedented challenges for the maintenance staffs. Greatly increased component sizes and load limits have placed tremendous responsibility on the machinery designer and the lubricant system manufacturer, to design and engineer systems that can be counted on to effectively lubricate and sustain these machines.

Fig. P&H 4100XPB Shovel

These machines are often as large as houses, with some booms extending nearly 300 feet. Depending on the type of machine, whether it is a shovel or a dragline or a drill, the machine's key components may be either electrohydraulic or electromechanical. For instance, the P&H 4100XPB shovel is driven solely by electric motors (no diesel engines) to turn the cab and operate the lift, while the Komatsu Demag Model H655S employs state-of-the-art electrohydraulic systems. Some draglines have huge hydraulic systems and electrically driven gears, with sump capacities into the thousands of gallons. The rolling stock, including bulldozers and large haul trucks, are powered by diesel engines. Many of the loading and hauling equipment designs incorporate hydraulic systems powered by diesel generators.

Fig. Komatsu Demag Model H655S

As one might expect, effective lubrication of these heavily loaded components is absolutely critical to efficient, reliable, economical operation of a mine site. As the machines have grown in size and sophistication, the manufacturers of lubricant products and delivery systems have had to work equally hard to match the new requirements.

There are two equally important aspects of lubrication effectiveness in the strip mine: the delivery system and the lubricant.

Lubrication Systems

Lubrication distribution systems have improved greatly over the years. The reasonable facsimile of what would have been used on the Bucyrus 24 dragline in the early 1900s. The lower funnel-shaped half of the unit would be threaded into a port in the bearing housing. The upper cylinder-shaped half would be full of multipurpose grease (MPG). Turning the upper cylinder clockwise would lower the cylinder onto the funnel-shaped portion, forcing (extruding) the lubricant stored in the reservoir of the cylinder portion through the funnel into the bearing. The lube tech would make routine lubrication checks throughout the machine. As required, these lubricators would be turned to ensure the lubricated components received adequate replenishment while in use.

The Lube Systems of Today

Today's centralized lubrication systems factory-installed on mining machinery have greatly minimized the need for lube techs in most cases. These computerized systems are now capable of dispensing the right product, in the right amount, to the right point, and at the right time. There are a number of centralized lubrication system manufacturers that offer a wide range of system types and designs to meet the needs of practically any mining application. Most lubrication systems consist of a grease pump, a motor to drive that pump and some sort of injector or valve to control or measure lubricant volume. Most utilize a programmable logic controller (PLC) to program the frequency of the lubrication replenishment cycle. As sophisticated as some of these systems are, maintenance and proper set-up is essential to ensure that their benefits are realized.

Upgrading

With the advancements in centralized lubrication systems, serious consideration should be taken when selecting a system or upgrading an existing system. Although the system's operation is generally quite simple and effective, performance limitations can have detrimental effects to lubricated components, such as the impact of changing lubricant viscosity. If a lubrication system is modified to accommodate more lube points or there is a need to change a lubricant product, it is strongly recommended that both the system manufacturer and lubricant manufacturer be consulted before such changes are made.

Lubricants

Lubricant manufacturers have continued to improve their products to meet the needs of bigger, faster machines. Although most lubricant suppliers are not lubrication system specialists, many have the resources to provide technical support, offering sound advice for selecting the products best-suited for the applications. The products commonly used in mining equipment can be divided into three groups: heavy-duty lubricating oils, such as EP oils for enclosed gear drives; multipurpose engine, circulating and hydraulic oils for engine, bearing lubrication and fluid power; and general purpose grease, for normal industrial bearing applications and specialized mining products. Walking draglines may require lubricants for the very large plain bearings that support the entire frame of the unit as it moves through the walking process.

These lubricants may have a high concentration of lubricating solids or soft metals dispersed into a stiff grease and delivered in small bags (for the walking mechanism without an automatic delivery system) just ahead of the peak loading area. This grease is referred to as a Walking Cam lubricant. An effort to reduce the number of lubricants on a machine has driven the development of multipurpose products designed to meet several different applications from a single lubrication system. The various components to be lubricated may include the open gears, guide rails, main table bearings and various smaller slides and bearings.

This presents a variety of challenges to tackle with one or two products. Specialty product manufacturers strive to meet the wide range of challenges with a single product. Despite the large volumes involved in filling a dragline during a lubricant change, given the product sophistication, some of these products may cost as much as $2.50 per pound. Management of the lubricant consumption to control costs becomes an integral part of the sourcing process.

Lubricant Management - Supply and Delivery

Vendors offer various options for purchase and delivery of these products, ranging from simply purchasing and managing the distribution of lubricants throughout the mine site, to programs offering on-board delivery, lubrication system set-up and long-term maintenance. Vendors may use several methods of delivery and replenishing lubrication systems and reservoirs. Today's service equipment, like the delivery truck is designed to meet the exacting mine delivery requirements. This may include a system of nozzles and hoses piped through filters for each type of lubricant as well as the necessary drive train to assure that the loaded vehicle can get out to the equipment under the worst of driving conditions.

Fig. Delivery Truck with Lubrication System

As with any industrial lubrication application, cleanliness must be first and foremost. Contaminated lubricants will dramatically reduce the life of the lubrication system components as well as the lubricated components. The dispensing equipment installed on the truck offers a completely self-contained, clean storage delivery environment, which helps to reduce the risk of contamination for all of the products that are delivered to the equipment.

Full-Service Supply Programs

Product-service combinations have been around the industry for years. Initially offered by specialty lubricant manufacturers, product-service combinations have proved valuable. Most lubricant manufacturers now offer some kind of service supporting their products. The ownership of full-service lubrication programs is completely taken over by the vendor, sometimes with the complete offering being purchased on a cost-per-hour program. Because many lubricant vendors resist consigning products to a mine, the cost-per-hour contract offers the next best option.

The purchase of the product is based on the hours of machinery operation. Sometimes these supply programs are provided to mines in conjunction with other full-service offerings. While this may sound ideal, vendors typically profit from selling less lubricant. Therefore, caution should be exercised when choosing this type of program-vendor combination. For the simple cost of this combination, the mines should receive a prenegotiated amount of monthly service along with the product. The service cost would be bundled into the cost of product.

These programs are advantageous to all when negotiated properly. As long as the level of service doesn't decline over the term of the contract, the benefits of the product-service combination can be considerable.

Lubrication Services Specialists

More recently, independent lubrication service consultants are becoming a viable alternative to the industry. Utilizing an independent consultant offers a mine the ability to purchase the product of choice based strictly on quality and product cost, but without any possible hidden costs of product-service combinations or cost-per-hour contracts.

The operators purchase the lube products for the equipment based on equipment criteria and purchase the service of a consultant based on experience and costs. This platform is a unique and upfront approach to product-service combinations. When considering an independent lubrication service consultant, check the individual's references supporting his/her abilities and knowledge.

STEAM SHOVEL

Fig. A derelict steam shovel in Alaska; major components visible include boiler, water tank, winch, main engine, boom, dipper stick, crowd engine, wheels and bucket

As the industry continues to evolve, expect to see continued evolution in all aspects of the industry and allied fields. With global positioning satellites (GPS) offering the potential for remotely operated equipment, computer systems taking lubrication systems to new levels of control, manufacturers continuing to meet the demands of an ever-changing and competitive industry, one thing should always remain the same. When it comes to the development and application of lubricating products, providing the cleanest possible environment, storing the products properly, reducing rehandling and applying the right product - in the right amount, in the right place, at the right time - will always be the necessary criteria, no matter how many times these practices are reinvented. A steam shovel is a large steam-powered excavating machine designed for lifting and moving material such as rock and soil. It is the earliest type of power shovel or excavator. They played a major role in public works in the 19th and early 20th century, being key to the construction of railroads and the Panama Canal. The development of simpler, cheaper diesel-powered shovels caused steam shovels to fall out of use in the 1930s.

HISTORY

Origins and development

Fig. A Link-Belt steam shovel crane, circa 1890.

The steam shovel was invented by William Otis, who received a patent for his design in 1839. The first machines were known as 'partial-swing', since the boom could not rotate through 360 degrees. They were built on a railway chassis, on which the boiler and movement engines were mounted. The shovel arm and driving engines were mounted at one end of the chassis, which accounts for the limited swing. Bogies with flanged wheels were fitted, and power was taken to the wheels by a chain drive to the axles. Temporary rail tracks were laid by workers where the shovel was expected to work, and repositioned as required.

Steam shovels became more popular in the latter half of the nineteenth century. Originally configured with chain hoists, the advent of steel cable in the 1870s allowed for easier rigging to the winches. Later machines were supplied with caterpillar tracks, obviating the need for rails. The full-swing, 360° revolving shovel was developed in England in 1884, and became the preferred format for these machines.

Growth and uses

Fig. A Marion steam shovel excavating thePanama Canal in 1908.

Fig. A steam shovel excavating for the San Diego and Arizona Railway line, circa 1919.

Expanding railway networks (in the US and the UK) fostered a demand for steam shovels. The extensive mileage of railways, and corresponding volume of material to be moved, forced the technological leap. As a result, steam shovels became commonplace.

American manufacturers included the Marion Steam Shovel Company, which was founded in 1884, Erie, P and H, and Bucyrus-ErieShovel Companies. The booming cities in North America used shovels to dig foundations and basements for the early skyscrapers. Perhaps the most famous application of steam shovels is the digging of the Panama Canal across the Isthmus of Panama. One hundred and two shovels worked in that decade-long dig. Of these, seventy-seven were built by Bucyrus; the remainder were Marion shovels.

These machines 'moved mountains' in their labours. The shovel crews would race to see who could move the most dirt. Steam shovels assisted mining operations: the iron mines of Minnesota, the copper mines of Chile and Montana, placer mines of the Klondike – all had earth-moving equipment. But it was with the burgeoning open-pit mines – first in Bingham Canyon, Utah – that shovels came into their own. The shovels systematically removed hillsides. As a result, steam shovels were used around the world from Australia to Russia to coal mines in China. Shovels were also used for construction, road and quarry work. Steam shovels came into their own in the 1920s with the publicly funded road building programs around North America. Thousands of miles of State Highways were built in this time period, together with new factories, such as Henry Ford's River Rouge Plant, and many docks, ports, buildings, and grain elevators.

Successors

During the 1930s steam shovels lost out to the simpler, cheaper diesel-powered excavating shovels that were the forerunners of those still in use today. Open-pit mines were electrified at this time. Only after the Second World War, with the advent of robust high-pressure hydraulic hoses, did the more versatile hydraulic backhoe shovels take pre-eminence over the cable-hoisting winch shovels.

Many steam shovels remained at work on the railways of developing nations until diesel engines supplanted them. Most have since been scrapped. Large, multi-ton mining shovels still use the cable-lift shovel arrangement. In the 1950s and 1960s Marion Shovel built massive stripping shovels for coal operations in the Eastern US. Shovels of note were the Marion 360, the Marion 5900, and the Marion 6360 – with a 180-cubic-yard (140 m3) bucket – while Bucyrus constructed one of the most famous monsters: the Big Brutus, the largest power shovel ever built and the largest still in existence.

The GEM of Egypt (GEM standing for "Giant Excavating Machine" and Egypt referring to the Egypt Valley in Belmont County, eastern Ohio where it was first put to use), which operated from 1967 to 1988, was of comparable size; it has since been dismantled. Although these big machines are still called steam shovels, they are more correctly known as power shovels since they use electricity to wind their winches.

OPERATION

A steam shovel consists of:

- A bucket, usually with a toothed edge, to dig into the earth
- A "dipper" or "dipper stick" connecting the bucket to the boom
- A "boom" mounted on the rotating platform, supporting the dipper and its control wires
- A boiler
- A water tank and coal bunker
- Steam engines and winches
- Operator's controls

- A rotating platform on a truck, on which everything is mounted
- Wheels (or sometimes caterpillar tracks or railroad wheels)
- A house (on the platform) to contain and protect 'the works'

The shovel has several individual operations: it can raise or luff the boom, rotate the house, or extend the dipper stick with the boom or crowd engine, and raise or lower the dipper stick.

When digging at a rock face, the operator simultaneously raises and extends the dipper stick to fill the bucket with material. When the bucket is full, the shovel is rotated to load a railway car or motor truck. The locking pin on the bucket flap is released and the load drops away. The operator lowers the dipper stick, the bucket mouth self-closes, the pin relocks automatically and the process repeats.

Steam shovels usually had a three-man crew: engineer, fireman and ground man. There was much jockeying to do to move shovels: rails and timber blocks to move; cables and block purchases to attach; chains and slings to rig; and so on. On soft ground, shovels used timber mats to help steady and level the ground. The early models were not self-propelled, rather they would use the boom to manoeuvre themselves.

STEAM SHOVEL MANUFACTURERS

Fig. A railroad shovel by Bucyrus; the boom rotates independently of the cab housing

North American manufacturers:

- Ball Engine Co.
- Bucyrus
- Erie
- Marion Steam Shovel Dredge Company
- Moore Speedcrane (later Manitowoc Cranes)
- Northwest Shovels
- Vulcan Iron Works

European manufacturers:

- Demag (Germany)
- Fiorentini (Italy)
- Lubecker
- Menck
- Newton & Chambers (UK)
- Orenstein and Koppel GmbH (Germany)
- Ruston & Hornsby (UK)

Power shovel/dragline manufacturers[edit]

- Bucyrus International
- Insley Manufacturing Co.
- Komatsu
- Lima Locomotive Works
- Link Belt
- Marion Power Shovel
- P&H Mining Equipment
- Priestman Bros (UK)
- Ransomes & Rapier (UK)
- Ruston-Bucyrus (UK)

PRESERVATION

Fig. Ruston-Bucyrus No. 4 (built 1931)

Fig. 1923 Bucyrus Model 50-B at the Nederland Mining Museum

Most steam shovels have been scrapped, although a few reside in industrial museums and private collections.

The Le Roy Marion

The world's largest steam shovel surviving intact is a 1906-built Marion machine, located in the small American town of Le Roy, New York. It was listed on the National Register of Historic Places in 2008.

Ruston Proctor Steam Navvy No 306

This machine was originally used at a chalk pit at Arlesey, in Bedfordshire, England. After the pit was closed, the steam navvy was simply abandoned and 'lost' as the pit became flooded with water.

By the mid-1970s, the area had become a local beauty spot, known as The Blue Lagoon (from chemicals from the quarry colouring the water), and after long periods of drought, the top of the rusty navvy could be seen protruding from the water. Ruston & Hornsby expert Ray Hooley heard of its existence, and organised the difficult task of rescuing it from the water-filled pit.

Hooley arranged for its complete restoration to working order by apprentices at the Ruston-Bucyrus works. Subsequently it passed into the care of the Museum of Lincolnshire Life, although it is not known whether it remains in operational condition. The navvy was recovered with hundreds of hours labour and free help from many organisations but today (2010) this Lincolnshire industrial monument stands unprotected with water allowed to pour down the boiler chimney. Many Lincolnshire residents and Ruston engineers wish to see this icon rescued.

1923 Bucyrus Model 50-B

Twenty-five Bucyrus Model 50-B steam shovels were sent to the Panama Canal to build bridges, roads, and drains and remove the huge quantities of soil and rock cut from the canal bed. All the shovels but one were scrapped at Panama. The survivor was shipped back to California and then brought to Denver. In the early 1950s it was transported to Rollinsville by Roy and Russell Durand, who operated it at the Lump Gulch Placer, six miles south of Nederland, Colorado, until 1978.

This steam shovel is one of two (the other at the Western Minnesota Steam Thresher's Reunion in Rollag, MN) remaining operational Bucyrus Model 50-Bs, and is preserved at the Nederland Mining Museum. Roots of Motive Power in Willits, CA has also acquired a 50-B and operates it for the public once a year at their Steam Festival in early September.

IN FICTION

Steam shovels have found literary fame with the children's classic Mike

Mulligan and His Steam Shovel. Another power shovelfeatures towards the climax of the children's book Are You My Mother? by P. D. Eastman – the little bird is returned to its nest by a "Snort" – but this is not necessarily a steam shovel.

Fig. Model of a steam shovel built fromMeccano and powered by a restored 1929 Meccano steam engine. (Model detailshere)

In the Thomas and Friends TV series, a steam shovel called Ned appears as a minor character. In the Australian children's TV series Mr. Squiggle, Bill the Steam Shovel provides comic relief and produces steam from his "nose" when he laughs. In an episode of Futurama, Bender is quoting as saying "When I grow up, I wanna be a steam shovel"

POWER SHOVEL

Fig. P & H 4100 XPB cable loading shovel.

A power shovel (also stripping shovel or front shovel or electric mining shovel) is a bucket-equipped machine, usually electrically powered, used for digging and loading earth or fragmented rock and for mineral extraction.

DESIGN

Shovels normally consist of a revolving deck with a power plant, driving and controlling mechanisms, usually a counterweight, and a front attachment, such as a crane ("boom") which supports a handle ("dipper" or "dipper stick") with a digger ("bucket") at the end.

"Dipper" is also sometimes used to refer to the handle and digger combined. The machinery is mounted on a base platform with tracks or wheels. Modern bucket capacities range from 8 m3 to nearly 80 m3.

Use

Fig. Shovel digging overburden.

Power shovels are used principally for excavation and removal of overburden in open-cut mining operations, though it may include loading of minerals, such as coal. They are the modern equivalent of steam shovels, and operate in a similar fashion.

Operation

The shovel operates using several main motions:

- Hoist - pulling the bucket up through the bank (*i.e.* the bank of material being dug)
- Crowd - moving the dipper handle out or in to control the depth of cut and when positioning to dump
- Swing - rotating the shovel between digging and dumping
- Propel - moving the shovel unit to different locations or dig positions

A shovel's work cycle, or digging cycle, consists of four phases:

- Digging
- Swinging

- Dumping
- Returning

The digging phase consists of crowding the dipper into the bank, hoisting the dipper to fill it, then retracting the full dipper from the bank. The swinging phase occurs once the dipper is clear of the bank both vertically and horizontally.

The operator controls the dipper through a planned swing path and dump height until it is suitably positioned over the haul unit (*e.g.* truck). Dumping involves opening the dipper door to dump the load, while maintaining the correct dump height. Returning is when the dipper swings back to the bank, and involves lowering the dipper into the tuck position to close the dipper door.

GIANT STRIPPING SHOVELS

Fig. Big Brutus.

In the 1950s with the demand for coal at a peak high and more coal companies turning to the cheaper method of strip mining, excavator manufacturers started offering a new super class of power shovels, commonly called giant stripping shovels.

Most were built between the 1950s and the 1970s. The world's first giant stripping shovel for the coal fields was the Marion 5760. Unofficially known to its crew and eastern Ohio residences alike as The Mountaineer, it was erected in 1955/56 near Cadiz, Ohio off of Interstate I-70. Larger models followed the successful 5760, culminating in the mid 60s with the gigantic 15,000 tonMarion 6360, nicknamed The Captain.

Ranking	Bucket Capacity (m^3/yd^3)	Operating weight (tons)	Type	Name	Service	Scrapped
	50/65	2,750	Marion 5760	The Mountaineer	1956	1988
	138/180	15,000	Marion 6360	The Captain	1965	1992
	107/140	9,350	Bucyrus Erie 3850B	River King	1964	1993
	99/130	6,850	Bucyrus Erie 1950B	GEM of Egypt	1967	1991
	96/125	9,338	Marion 5960	Big Digger	1969	1989
	88/115	6,950	Bucyrus Erie 3850B	The Big Hog	1962	1985 (Accident)
	80/105	7,200	Bucyrus Erie 1950B	The Silver Spade	1965	2007
	69/90	5,220	Bucyrus-Erie 1850-B	Big Brutus	1962	Preserved as a National Landmark

EXCAVATOR

Fig. A typical modern excavator: a CAT 325C, fitted with quick coupler and tilting bucket.

Excavators are heavy construction equipment consisting of a boom, stick, bucket and cab on a rotating platform (known as the "house"). The house sits atop an undercarriage with tracks or wheels. A cable-operated excavator uses winches and steel ropes to accomplish the movements. They are a natural progression from the steam shovels and often called power shovels. All movement and functions of a hydraulic excavator are accomplished through the use of hydraulic fluid, with hydraulic cylinders and hydraulic motors. Due to the linear actuation of hydraulic cylinders, their mode of operation is fundamentally different from cable-operated excavators.

Terminology

Excavators are also called diggers, JCBs (a proprietary name, in an example of a generic trademark), mechanical shovels, or 360-degree excavators (sometimes abbreviated simply to 360). Tracked excavators are sometimes called "trackhoes" by analogy to the backhoe. In the UK, wheeled excavators are sometimes known as "rubber ducks."

Usage

Excavators are used in many ways:

- Digging of trenches, holes, foundations
- Material handling
- Brush cutting with hydraulic attachments
- Forestry work
- Forestry mulching
- Demolition
- General grading/landscaping
- Heavy lift, *e.g.* lifting and placing of pipes
- Mining, especially, but not only open-pit mining
- River dredging
- Driving piles, in conjunction with a pile driver

Fig. An old excavator under the Northwest (now Terex) name at the Pageant of Steam grounds

Fig. Excavator demolishing a house. Note the hydraulicthumb

Fig. Link-Belt excavator trenching

Fig. CAT 5230 in coal mining operation

Fig. A L&T Komatsu Excavator seen in India

CONFIGURATIONS

Excavators come in a wide variety of sizes. The smaller ones are called mini or compact excavators. Caterpillar's smallest mini-excavator weighs 2,060 pounds (930 kg) and has 13 hp; their largest model is the largest excavator available (a record previously held by the Orenstein & Koppel RH400) the CAT 6090, it weighs in excess of 2,160,510 pounds (979,990 kg), has 4500 hp and has a bucket size of around 52.0 m^3 depending on bucket fitted.

Engines in excavators drive hydraulic pumps; there are usually 3 pumps: the two main pumps are for supplying oil at high pressure (up to 5000 psi) for the rams, swing motor, track motors, and accessories, and the third is a lower pressure (700 psi) pump for Pilot Control, this circuit used for the control of the spool valves, this allows for a reduced effort required when operating the controls.

The two main sections of an excavator are the undercarriage and the house. The undercarriage includes the blade (if fitted), tracks, track frame, and final drives, which have a hydraulic motor and gearing providing the drive to the individual tracks, and the house includes the operator cab, counterweight, engine, fuel and hydraulic oil tanks. The house attaches to the undercarriage by way of a center pin. High pressure oil is supplied to the tracks' hydraulic motors through a hydraulic swivel at the axis of the pin, allowing the machine to slew 360° unhindered.

The main boom attaches to the house, and can be one of several different configurations:

- Most are mono booms: these have no movement apart from straight up and down.
- Some others have a knuckle boom which can also move left and right in line with the machine.
- Another option is a hinge at the base of the boom allowing it to hydraulically pivot up to 180° independent to the house; however, this is generally available only tocompact excavators.
- There are also triple-articulated booms (TAB).

Attached to the end of the boom is the stick (or dipper arm). The stick provides the digging force needed to pull the bucket through the ground. The stick length is optional depending whether reach (longer stick) or break-out power (shorter stick) is required.

On the end of the stick is usually a bucket. A wide, large capacity (mud) bucket with a straight cutting edge is used for cleanup and levelling or where the material to be dug is soft, and teeth are not required. A general purpose (GP) bucket is generally smaller, stronger, and has hardened side cutters and teeth used to break through hard ground and rocks. Buckets have numerous shapes and sizes for various applications. There are also many other attachments which are available to be attached to the excavator for boring, ripping, crushing, cutting, lifting, etc.

Before the 1990s, all excavators had a long or conventional counterweight that hung off the rear of the machine to provide more digging force and lifting capacity. This became a nuisance when working in confined areas. In 1993 Yanmar launched the world's first Zero Tail Swing excavator, which allows the counterweight to stay inside the width of the tracks as it slews, thus being safer and more user friendly when used in a confined space. This type of machine is now widely used throughout the world.

There are two main types of "Control" configuration generally use in excavators to control the boom and bucket, both of which spread the four main digging controls between two x-y joysticks. This allows a skilled operator to control all four functions simultaneously. The most popular configuration in the US is the SAE controls configuration while in other parts of the world, the ISO control configuration is more common. Some manufacturers such as Takeuchi have switches that allow the operator to select which control configuration to use.

Excavator attachments

Hydraulic excavator capabilities have expanded far beyond excavation tasks with buckets. With the advent of hydraulic-powered attachments such as a breaker, a grapple or an auger, the excavator is frequently used in many applications other than excavation. Many excavators feature a quick coupler for simplified attachment mounting, increasing the machine's utilization on the jobsite. Excavators are usually employed together with loaders and bulldozers. Most wheeled, compact and some medium-sized (11 to 18-tonne) excavators have a backfill (or dozer) blade. This is a horizontal bulldozer-like blade attached to the undercarriage and is used for levelling and pushing removed material back into a hole.

DRAGLINE EXCAVATOR

Fig. Page 725 dragline at Turkish Coal Association-Yenikoy Open Pit Coal Mine, Milas

A dragline excavator is a piece of heavy equipment used in civil engineering and surface mining. Draglines fall into two broad categories: those that are based on standard, lifting cranes and the heavy units which have to be built on-site. Most crawler cranes with an added winch drum on the front can act as a dragline.

These units (like other cranes) are designed to be dismantled and transported over the road on flatbed trailers. Draglines used in civil engineering are almost always of this smaller, crane type. These are used for road, port construction, pond and canal dredging, and as piledriving rigs. These types are built by crane manufacturers such as Linkbelt and Hyster. The much larger type which is built on site is commonly used in strip-mining operations to remove overburden above coal and more recently for tar-sand mining. The largest heavy draglines are among the largest mobile land machines ever built. The smallest and most common of the heavy type weigh around 8,000 tons while the largest built weighed around 13,000 tons.

A dragline bucket system consists of a large bucket which is suspended from a boom (a large truss-like structure) with wire ropes. The bucket is manoeuvered by means of a number of ropes and chains. The hoist rope, powered by large diesel orelectric motors, supports the bucket and hoist-coupler assembly from the boom. The dragrope is used to draw the bucket assembly horizontally. By skillful manoeuvre of the hoist and the dragropes the bucket is controlled for various operations. A schematic of a large dragline bucket system is shown below.

HISTORY

The dragline was invented in 1904 by John W. Page (as a partner of the firm Page & Schnable Contracting) for use digging the Chicago Canal. By 1912, Page realized that building draglines was more lucrative than contracting so he created the Page Engineering Company to build draglines. Page built its first crude walking dragline in 1923.

These used legs operated by rack and pinion on a separate frame that lifted the crane. The body was then pulled forward by chain on a roller track and then lowered again. Page developed the first diesel engines exclusively for dragline application in 1924. Page also invented the arched dragline bucket, a design still commonly used today by draglines from many other manufacturers, and in the 1960s pioneered an archless bucket design.

With its walking mechanism badly behind Monighan, Page updated the mechanism to an eccentric drive in 1935. This much improved mechanism gave a proper elliptical motion and was used until 1988. Page modernized its draglines further with the 700 series in 1954. Page's largest dragline was the Model 757 delivered to the Obed Mine near Hinton, Alberta in 1983. It featured a 75-yard bucket on a 298-foot boom and an operating weight of 4,500 tons. In 1988, Harnischfeger Corporation (P&H Mining Equipment) purchased Page

Engineering Company. Harnischfeger Corporation was established as P&H Mining in 1884 by Alonzo Pawling and Henry Harnischfeger. In 1914, P&H introduced the world's first gasoline engine powered dragline. In 1988, Page was acquired by Harnischfeger which makes the P&H line of shovels, draglines, and cranes. P&H's largest dragline is the 9030C with a 160-yard bucket and up to a 425-foot boom.

In 1907, Monighan's Machine Works of Chicago became interested in manufacturing draglines when local contractor John W. Page placed an order for hoisting machinery to install a dragline. In 1908, Monighan changed its name to the Monighan Machine Company. In 1913, a Monighan engineer named Oscar Martinson invented the first walking mechanism for a dragline. The device, known as the Martinson Tractor, was installed on a Monighan dragline, creating the first walking dragline. This gave Monighan a significant advantage over other draglines and the company prospered. The cam mechanism was further improved in 1925 by eliminating the drag chains for the shoes and changing to a cam wheel running in an oval track. This gave the shoe a proper elliptical motion. The first dragline using the new mechanism was the 3-W available in 1926. So popular were these machines that the name Monighan became a generic term for dragline. In the early 1930s, Bucyrus-Erie began purchasing shares of Monighan stock with Monighan's approval. Bucyrus purchased a controlling interest and the joint company became known as Bucyrus-Monighan until the formal merger in 1946.

Ransomes & Rapier was founded in 1869 by four engineers to build railway equipment and other heavy works. In 1914 they started building two small Steam shovels as a result of a customer request. The rope operated crowd system they built for this was patented and later sold to Bucyrus. After WWI, demand for excavators increased and in 1924 they reached an agreement to build Marion draglines from 1 to 8 cubic yards capacity. In 1927, they built Type-7 1-yard and Type-460 1.5-yard models. The deal to build Marion machines ended in 1936. R&R began building their own designs with the Type-4120 followed by the 4140 of 3.5 cubic yards. In 1958 the Ramsomes & Rapier division was sold to Newton, Chambers & Co. Ltd of Sheffield, which was combined with their NCK Crane & Excavator division. This became NCK-Rapier. The walking dragline division of NCK-Rapier was acquired by Bucyrus in 1988.

The Marion Power Shovel Company (established in 1880) built its first walking dragline with a simple single-crank mechanism in 1939. Its largest dragline was the 8950 sold to Amax Coal Company in 1973. It featured a 150-cubic yard bucket on a 310-foot boom and weighed 7,300 tons. Marion was acquired by Bucyrus in 1997. Bucyrus International entered the dragline market in 1910 with the purchase of manufacturing rights for the Heyworth-Newman dragline excavator. Their "Class 14" dragline was introduced in 1911 as the first crawler mounted dragline. In 1912 Bucyrus helped pioneer the use of electricity as a power source for large stripping shovels and draglines used in mining. An Italian company, Fiorentini, produced dragline excavators from 1919 licensed by Bucyrus. After the merger with Monighan in 1946, Bucyrus began producing much larger machines using the Monighan walking mechanism such as the 800 ton 650-B which used a 15-yard bucket. Bucyrus' largest dragline was Big Muskiebuilt for the Ohio Coal Company in 1969. This machine featured a 220-yard bucket on a 450-foot boom and weighed 14,500 tons. Bucyrus was itself acquired by heavy equipment and diesel engine maker, Caterpillar, in 2011. Caterpillar's largest dragline is the 8750 with a 169-yard bucket, 435-foot boom, and 8,350 ton weight.

The market for draglines began shrinking rapidly after the boom of the 1960s and 1970s which led to more mergers. P&H's acquisition of Page in 1988 along with Bucyrus' acquisition of Ransomes & Rapier in 1988 and Marion in 1997 cut the number of worldwide suppliers of heavy draglines by more than half. Today, P&H and Caterpillar are the only remaining manufacturers of large draglines.

Other Manufacturers

Heavy Engineering Corporation Limited was the first Indian company to manufacture a walking Dragline of 31-yard bucket capacity. HEC makes up to a 44-yard bucket. For comparison, this would be comparable to Caterpillar's Small Draglines 8000 series with a 42-yard bucket. HEC has supplied slightly over a dozen to the Indian Mining industry.

OPERATION

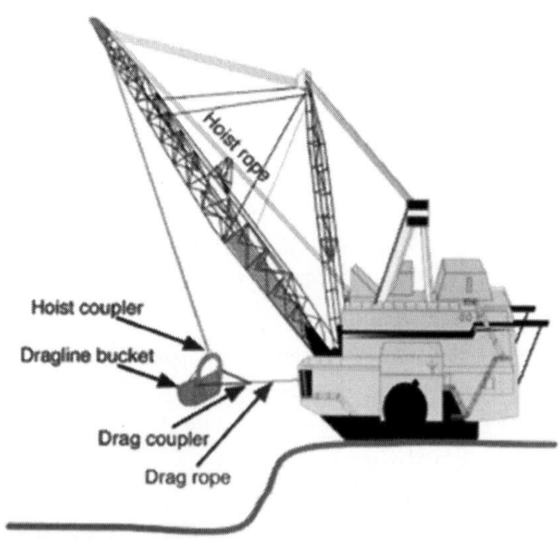

Fig. Dragline excavator.

In a typical cycle of excavation, the bucket is positioned above the material to be excavated. The bucket is then lowered and the dragrope is then drawn so that the bucket is dragged along the surface of the material. The bucket is then lifted by using the hoist rope. A swing operation is then performed to move the bucket to the place where the material is to be dumped. The dragrope is then released causing the bucket to tilt and empty. This is called a dump operation.

On crane-type draglines, the bucket can also be 'thrown' by winding up to the jib and then releasing a clutch on the drag cable. This would then swing the bucket like a pendulum. Once the bucket had passed the vertical, the hoist cable would be released thus throwing the bucket. On smaller draglines, a skilled operator could make the bucket land about one-half the length of the jib further away than if it had just been dropped. On larger draglines, this is not a common practice.

Draglines have different cutting sequences. The first is the side cast method using offset benches; this involves throwing the overburden sideways onto blasted material to make a bench. The second is a key pass. This pass cuts a key at the toe of the new highwall and also shifts the bench further towards the low-wall. This may also require a chop pass if the wall is blocky. A chop pass involves the bucket being dropped down onto an angled highwall to scale the surface. The next sequence is the slowest operation, the blocks pass. However, this pass moves most of the material. It involves using the key to access to bottom of the material to lift it up to spoil or to an elevated bench level. The final cut if required is a pull back, pulling material back further to the low-wall side.

DRAGLINES IN MINING

Fig. Dragline at the Curragh Coal Mine

A large dragline system used in the open pit mining industry costs approximately US$50–100 million. A typical bucket has a volume ranging from 40 to 80 cubic yards (30 to 60 cubic metres), though extremely large buckets have ranged up to 168 cubic metres (5,900 cu ft). The length of the boom ranges from 45 to 100 metres (148 to 330 ft). In a single cycle, it can move up to 450 metric tonnes of material. Most mining draglines are not diesel-powered like most other mining equipment. Their power consumption is so great that they have a direct connection to the high-voltage grid at voltages of between 6.6 and 22 kV. A typical dragline, with a 55-cubic-metre bucket, can use up to 6 megawatts during normal digging operations. Because of this, many (possibly apocryphal) stories have been told about the blackout-causing effects of mining draglines. For instance, there is a long-lived story that, back in the 1970s, if all seven draglines at Peak Downs Mine (a very large BHP coal mine in central Queensland, Australia) turned simultaneously, they would black out all of North Queensland. However even now, if they have been shut down, they are always restarted one at a time due to the immense power requirements of startup.

In all but the smallest of draglines, movement is accomplished by "walking" using feet or pontoons, as caterpillar tracks place too much pressure on the ground, and have great difficulty under the immense weight of the dragline. Maximum speed is only at most a few metres per minute,since the feet must be repositioned for each step. If travelling medium distances, (about 30–100 km), a special dragline carrier can be brought in to transport the dragline. Above that distance, disassembly is generally required. But mining draglines due to their reach can work a large area from one position and do not need to constantly move along the face like smaller machines.

Limitations

The primary limitations of draglines are their boom height and boom length, which limits where the dragline can dump the waste material. Another primary limitation is their dig depth, which is limited by the length of rope the dragline can utilize. Inherent with their construction, a dragline is most efficient excavating material below the level of their base. While a dragline can dig above itself, it does so inefficiently and is not suitable to load piled up material (as a rope shovel or wheel loader can). Despite their limitations, and their extremely high capital cost, draglines remain popular with many mines, due to their reliability, and extremely low waste removal cost.

Examples

The coal mining dragline known as Big Muskie, owned by the Central Ohio Coal Company (a division of American Electric Power), was the world's largest mobile earth-moving machine, weighing nearly 13,000 metric tons and standing nearly 22 stories tall. It operated in Muskingum County, in the U.S. state of Ohio from 1969 to 1991, and was powered by 13,800 volts of electricity. It was scrapped in 1999. The British firm of Ransomes & Rapier produced a few large (1400-1800 ton) excavators, the largest in Europe at the time (1960s).

Power was from internal combustion engines driving electric generators. One, named SUNDEW, was used in a quarry from 1957 to 1974. After its working life at the first site in Rutland was finished it walked 13 miles to a new life at Corby; the walk took 9 weeks.

Fig. The Walking Mechanism on a preservedBucyrus-Erie 1150 dragline in the UK

Smaller draglines were also commonly used before hydraulic excavators came into common use, the smaller draglines are now rarely used other than on river and gravel pit works. The small machines were of a mechanical drive with clutches. Firms such as Ruston and Bucyrus made models such as the RB10 which were popular for small building works and drainage work. Several of these can still be seen in the English Fens of Cambridgeshire, Lincolnshire and parts of Norfolk. Ruston's are a company also associated with drainage pumping engines. Electric drive systems were only used on the larger mining machines, most modern machines use a diesel-hydraulic drive, as machines are seldom in one location long enough to justify the cost of installing a substation and supply cables.

TECHNOLOGICAL ADVANCES

Draglines, unlike most equipment used in earth-moving, have remained relatively unchanged in design and control systems for almost 100 years. Over the last few years, some advances in dragline systems and methodologies have occurred.

Automation

Researchers at CSIRO in Australia have a long-term research project into automating draglines. Mining automation teams at QCAT, a CSIRO division; have been developing the automation technology since 1994. Automated systems include cruise control and Digital Terrain Mapping. Working solutions include the proof-of-concept dragline swing cruise control on a Tarong BE1370.

Simulation software

Since draglines are typically large, complicated and very expensive, training new operators can be a tricky process. In the same way that flight simulators have developed to train pilots, mining simulator software has been developed to assist new operators in learning how to control the machines.

UDD

UDD stands for Universal-Dig-Dump. It represents the first fundamental change to draglines for almost a century, since the invention of the 'miracle hitch'. Instead of using two ropes (the hoist rope and the drag rope) to manipulate the bucket, a UDD machine uses three ropes, two hoist and one drag. This allows the dragline operator to have much greater selectivity in when to pick up the bucket, and in how the bucket may be dumped.

UDD machines generally have higher productivity than a standard dragline, but often have greater mechanical issues. Within the mining industry, there is still much debate as to whether UDD improvements justify their costs.

BUCKET-WHEEL EXCAVATOR

Fig. Bucket wheel excavator in Ferropolis, Germany

Bucket-wheel excavators (BWEs) are heavy equipment used in surface mining. The primary function of BWEs is to act as a continuous digging machine in large-scale open pit mining operations. What sets BWEs apart from other large-scale mining equipment, such as bucket chain excavators, is their use of a large wheel consisting of a continuous pattern of buckets used to scoop material as the wheel turns.

They are among the largest vehicles ever constructed, and the biggest bucket-wheel excavator ever built, Bagger 293, is the largest terrestrial (land) vehicle in human history according to the Guinness Book of World Records.

HISTORY

Fig. Bucket-wheel excavator in the open-pit mining Garzweiler

Bucket-wheel excavators have been used in mining for the past century, with some of the first being manufactured in the 1920s.[1]They are used in conjunction with many other pieces of mining machinery (conveyor belts, spreaders, crushing stations, heap-leach systems, etc.) to move and mine massive amounts of overburden (waste). While the overall concepts that go into a BWE have not changed much, their size has grown drastically since the end of World War Two.

In the 1950s two German mining firms ordered the world's first extremely large BWEs, and had three BWEs built for mining lignite near Cologne, Germany. The German BWEs had a wheel of over 52 feet (16 m) in diameter, weighed 5,500 short tons (5,000 t) and was over 600 feet (180 m) long, with eighteen crawler units for movement and could cut a swath of over 600 feet (180 m) feet at one time

BWEs built since the 1990s, such as the Bagger 293, have reached sizes as large as 96 metres (314.9 feet) tall, 225 metres (738.2 feet) long, and as heavy as 14,200 metric tons (31.3 million lb). The bucket-wheel itself can be over 70 feet in diameter with as many as 20 buckets, each of which can hold over 15 cubic metres of material. BWEs have also advanced with respect to the extreme conditions in which they are now capable of operating. Many BWEs have been designed to operate in climates with temperatures as low as -45°C (-49°F). Developers are now moving their focus toward automation and the use of electrical power.

STRUCTURE

A bucket wheel excavator (BWE) consists of a superstructure to which several more components are fixed. The bucket wheel from which the machines get their name is a large, round wheel with a configuration of scoops which is fixed to a boom and is capable of rotating.

Material picked up by the cutting wheel is transferred back along the boom. In early cell-type bucket wheels, the material was transferred through a chute leading from each bucket, while newer cell-less and semi-cell designs use a stationary chute through which all of the buckets discharge. A discharge boom receives material through the superstructure from the cutting boom and carries it away from the machine, frequently to an external conveyor system.

A counterweight boom balances the cutting boom and is cantilevered either on the lower part of the superstructure (in the case of compact BWEs) or the upper part (in the case of mid-size C-frame BWEs). In the larger BWEs, all three booms are supported by cables running across towers at the top of the superstructure.

Beneath the superstructure lay the movement systems. On older models these would be rails for the machine to travel along, but newer BWEs are frequently equipped withcrawlers, which grant them increased flexibility of motion. To allow it to complete its duties, the superstructure of a BWE is capable of rotating about a vertical axis (slewing). The cutting boom can be tilted up and down (hoisting). The speeds of these operations are on the orders of 30 m/min and 5 m/min, respectively. Slewing is driven by large gears, while hoisting generally makes use of a cable system.

Size

The scale of BWEs varies drastically and is dependent on the intended application. Compact BWEs designed by ThyssenKrupp may have boom lengths as small as 6m, weigh 50 tons, and move 100 fm3/hr of earth. Their larger models reach boom lengths of 80m, weigh 13,000 tons, and move 12,500 fm3/hr. The largest BWE ever constructed is TAKRAF's Bagger 293, which weighs 14,200 (metric) tons and is capable of moving 240,000 cubic metres of overburden every day. Excavations of 380,000 cubic metres per day have been recorded. The BWEs used in the United States tend to be smaller than those constructed in Germany.

Bucket chain excavators

Bucket chain excavators (BCEs) are similar in structure and function to BWEs. However, instead of the buckets being placed in a ring, they are strung out in a manner reminiscent of a trencher. They remove material from below their plane of movement, which is useful if the pit floor is unstable or underwater. TAKRAF's BCEs travel on rails rather than crawlers.

OPERATION

BWEs are used for continuous overburden removal in surface mining applications. They use their cutting wheels to strip away a section of earth (the working block) dictated by the size of the excavator. Through hoisting, the working block can include area both above and below the level of the machine (the bench level). By slewing, the excavator can reach through a horizontal range.

The overburden is then delivered to the discharge boom, which transfers the cut earth to another machine for transfer to a spreader. This may be a fixed belt conveyor system or a mobile conveyor with crawlers similar to those found on the BWE. Mobile conveyors permanently attached to the excavator takes the burden of directing the material off of the operator. The overburden can also be transferred directly to cross-pit Spreader, which reaches across the pit and scatters overburden at the dumping ground.

Automation

Automation of the BWEs requires integrating many sensors and electrical components such as GPS, data acquisition systems, and online monitoring capabilities. The goal of these systems is to take away some of the work from the operators in order to achieve higher mining speeds. Project managers and operators are now able to track crucial data regarding the BWEs and other machinery in the mining operations via the Internet. Sensors can detect how much material is being scooped onto the conveyor belt, and the automation system can then vary the speed on the conveyor belts in order to feed a continuous amount of material.

APPLICATIONS

Bucket wheel excavators and bucket chain excavators take jobs that were previously accomplished by rope shovels and draglines. They have been replaced in most applications by hydraulic excavators, but still remain in use for very large-scale operations, where they can be used for the transfer of loose materials or the excavation of soft to semi-hard overburden.

Lignite mining

The primary application of BWEs is in lignite (brown coal) mining, where they are used for soft rock overburden removal in the absence of blasting. They are useful in this capacity for their ability to continuously deliver large volumes of materials to processors, which is especially important given the continuous demand for lignite.

Because of the great demand for lignite, lignite mining has also been one of the areas of greatest development for BWEs. The additions of automated systems and greater manoeuvrability, as well as components designed for the specific application, have increased the reliability and efficiency with which BWEs deliver materials.

Materials handling

Bucket wheel technology is used extensively in bulk materials handling. Bucket wheel reclaimers are used to pick up material that has been positioned by a stacker for transport to a processing plant. Stacker/reclaimers, which combine tasks to reduce the number of required machines, also use bucket wheels to carry out their tasks. In shipyards, bucket wheels are used for the continuous loading and unloading of ships, where they pick up material from the yard for transfer to the delivery system. Bucket chains can be used to unload material from a ship's hold. TAKRAF's continuous ship unloader is capable of removing to 95% of the material from a ship's hold, owing to a flexibly-configured digging attachment.

Heap leaching

An extension of their other uses, BWEs are used in heap leaching processes. Heap leaching entails of constructing stacks of crushed ore, through which a solvent is passed to extract valuable materials. The construction and removal of the heaps are an obvious application of stacking and reclaiming technology.

MANUFACTURERS AND MARKET

Few companies are willing or able to manufacture the massive, expensive gears required for BWEs. Unex, Czech Republic, still has the original casting forms, and is still able to manufacture BWEs. However, these machines were built to last indefinitely under continuous heavy use, and strip mining is now ecologically unpopular, so there is little demand for new machines. The manufacturers of BWEs and similar mining systems now receive some revenue from maintenance and refurbishing projects, but also produce large steel parts for other purposes. Current use of bucket-wheel excavators is mainly focused in the area of lignite (brown coal) mining for the production of electricity, mostly in Germany and East/Southeastern Europe. Unex has also made a BWE for extraction of diamonds from the Siberian permafrost. There are other manufacturers as well, Sandvik Mining and Construction, can offer systems for open pit mining.

In culture

A bucket-wheel excavator is prominently featured in the 2011 film "Ghost Rider: Spirit of Vengeance" as the temporary steed of the titular character. As a result it becomes infused with the blazing flames that rage from whatever vehicle the Ghost Rider occupies.

DREDGING

Dredging is an excavation activity or operation usually carried out at least partly underwater, in shallow seas or fresh water areas with the purpose of gathering up bottom sediments and disposing of them at a different location. This technique is often used to keep waterways navigable. It is also used as a way to replenish sand on some public beaches, where sand has been lost because of coastal erosion. Dredging is also used as a technique for fishing for certain species of edible clams and crabs, see fishing dredge.

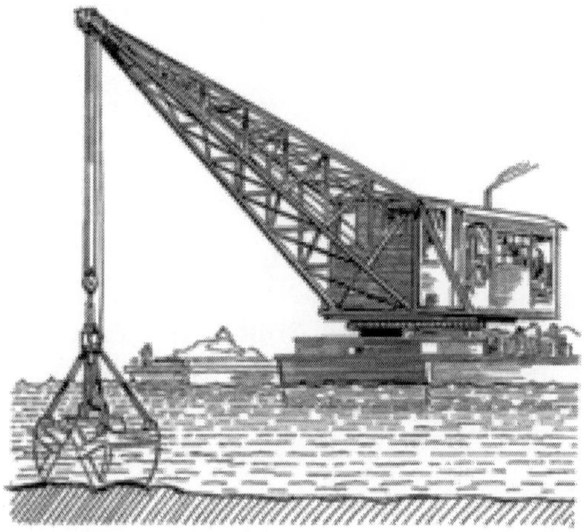

Fig. A grab dredge

A dredger (or "dredge" as is the general usage in the Americas) is any device, machine, or vessel that is used to excavate and remove material from the bottom of a body of water. For example, a scoop attached to the end of a rope or pole by which a man can draw sediments up from the bottom of a pond is a dredger. Developing this idea further, a motorized crane equipped with a drag bucket or clamshell (grabber) that is used to scoop material from the bottom of a body of water is also a dredger. The crane could be located on the bank, or perhaps mounted on a barge. If the crane is mounted on a barge, the entire vessel is referred to as a dredger.

The process of dredging creates spoils (excess material), which are carried away from the dredged area. Dredging can produce materials for land reclamation or other purposes (usually construction-related), and has also historically played a significant role in gold mining. Dredging can create disturbance in aquatic ecosystems, often with adverse impacts.

Uses

- Capital: dredging carried out to create a new harbour, berth or waterway, or to deepen existing facilities in order to allow larger ships access. Because capital works usually involve hard material or high-volume works, the work is usually done using a cutter suction dredge or large trailing suction hopper dredge; but for rock works, drilling and blasting along with mechanical excavation may be used.
- Preparatory: work and excavation for future bridges, piers or docks/ wharves, often connected with foundation work.
- Maintenance: dredging to deepen or maintain navigable waterways or channels which are threatened to become silted with the passage

of time, due to sedimented sand and mud, possibly making them too shallow for navigation. This is often carried out with a trailing suction hopper dredge. Most dredging is for this purpose, and it may also be done to maintain the holding capacity of reservoirs or lakes.

- Land reclamation: dredging to mine sand, clay or rock from the seabed and using it to construct new land elsewhere. This is typically performed by a cutter-suction dredge or trailing suction hopper dredge. The material may also be used for flood or erosion control.
- Beach nourishment: mining sand offshore and placing on a beach to replace sand eroded by storms or wave action. This is done to enhance the recreational and protective function of the beaches, which can be eroded by human activity or by storms. This is typically performed by a cutter-suction dredge or trailing suction hopper dredge.
- Harvesting materials: dredging sediment for elements like gold, diamonds or other valuable trace substances.
- Seabed mining: a possible future use, recovering natural metal ore nodules from the sea's abyssal plains.
- Construction materials: dredging sand and gravels from offshore licensed areas for use in construction industry, principally for use in concrete. Very specialist industry focused in NW Europe using specialized trailing suction hopper dredgers self discharging dry cargo ashore.
- Anti-eutrophication: Dredging is an expensive option for the remediation of eutrophied (or de-oxygenated) water bodies. However, as artificially elevated phosphorus levels in the sediment aggravate the eutrophication process, controlled sediment removal is occasionally the only option for the reclamation of still waters.
- Contaminant remediation: to reclaim areas affected by chemical spills, storm water surges (with urban runoff), and other soil contaminations. Disposal becomes a proportionally large factor in these operations.
- Removing trash and debris: often done in combination with maintenance dredging, this process removes non-natural matter from the bottoms of rivers and canals and harbors.
- Flood prevention: this can help to increase channel depth and therefore increase a channel's capacity for carrying water.
- Peat extraction: in former times, so-called dredging poles or dredge hauls were used on the back of small boats to manually dredge the beds of peat-moor waterways before extracting the peat for use as a fuel. This tradition has now become more or less obsolete and the tools used to do this have also changed significantly.

- Oyster dredging or harvesting: in Louisiana and other states with salt water estuaries that can sustain bottom oyster beds. A heavy metal retangular scoop device is towed astern of a moving boat with a chain bridle attached to a cable and winch which scoops up oysters as it drags along the bottom. The device is periodically hauled aboard and the oysters in it are sorted and bagged for shipment to an oyster processing facility.
- As a hobby: hobbyists examine their dredged matter to pick out items of potential value, similar to the hobby of metal detecting, or the hobby form of dumpster diving, on land.

Relevance

Without the many and almost non-stop dredging operations world wide, much of the world's commerce would be impaired, often within a few months, since much of world's goods travel by ship, and need to access harbours or seas via channels. Recreational boating also would be constrained to the smallest vessels.

The majority of marinedredging operations (and the disposal of the dredged material) will require that appropriate licences are obtained from the relevant regulatory authorities, and dredging is usually carried out by (or for) harbour companies or corresponding government agencies.

TYPES OF DREDGING VESSELS

Suction

Fig. The Geopotes 14 lifting its boom on a canal in The Netherlands. (geopotes is Greek for "that which drinks earth")

These operate by sucking through a long tube, like some vacuum cleaners but on a larger scale. A plain suction dredger has no tool at the end of the suction pipe to disturb the material. This is often the most commonly used form of dredging.[citation needed]

Trailing suction

A trailing suction hopper dredger (TSHD) trails its suction pipe when working, and loads the dredge spoil into one or more hoppers in the vessel. When the hoppers are full, the TSHD sails to a disposal area and either dumps the material through doors in the hull or pumps the material out of the hoppers.

Some dredges also self-offload using drag buckets and conveyors. The largest trailing suction hopper dredger in the world are currently Jan De Nul's Cristobal Colon (launched 4 July 2008) and its sister ship Leiv Eriksson (launched 4 September 2009). Main design specs for the Cristobal Colon and the Leiv Eriksson are: 46,000 cubic metre hopper and a design dredging depth of 155 m. Next largest is HAM 318 (Van Oord) with its 37,293 cubic metre hopper and a maximum dredging depth of 101 m.

Cutter-suction

A cutter-suction dredger's (CSD) suction tube has a cutting mechanism at the suction inlet. The cutting mechanism loosens the bed material and transports it to the suction mouth. The dredged material is usually sucked up by a wear-resistant centrifugal pump and discharged either through a pipe line or to a barge.

Cutter-suction dredgers are most often used in geological areas consisting of hard surface materials (for example gravel deposits or surface bedrock) where a standard suction dredger would be ineffective. In recent years, dredgers with more powerful cutters have been built in order to excavate harder rock without the need for blasting. The two largest cutter suction dredgers in the world are currently (as at August 2009) DEME's D'Artagnan (28,200 kW total installed power) and Jan De Nul's J.F.J. DeNul(27,240 kW). both built by IHC Merwede.

Auger suction

This process functions like a cutter suction dredger, but the cutting tool is a rotating Archimedean screw set at right angles to the suction pipe. The first widely used auger dredges were designed by Mud Cat Dredges in the 1980s which was run by National Car Rental, but is now a Division of Ellicott Dredges. In 1996, IMS Dredges introduced a self-propelled version of the auger dredge that allows the system to propel itself without the use of anchors cables. During the 1980s and 1990s auger dredges were primarily used for sludge removal applications from waste water treatment plants. Today, auger dredges are used for a wider variety of applications including river maintenance and sand mining. The most common auger dredge on the global market today is the Versi-Dredge. The turbidity shroud on auger dredge systems creates a strong suction vacuum and creates much less turbidity than conical (basket) type cutterheads and that is why they are preferred for environmental type applications.

The vacuum created by the shroud and the ability to convey material to the pump faster makes auger dredge systems more productive than similar sized conical (basket) type cutterhead dredges.

Jet-lift

These use the Venturi effect of a concentrated high-speed stream of water to pull the nearby water, together with bed material, into a pipe.

Air-lift

An airlift is a type of small suction dredge. It is sometimes used like other dredges. At other times, an airlift is used, handheld underwater by a diver. It works by blowing air into the pipe, and that air, being lighter than water, rises inside the pipe, dragging water with it.

Bucket

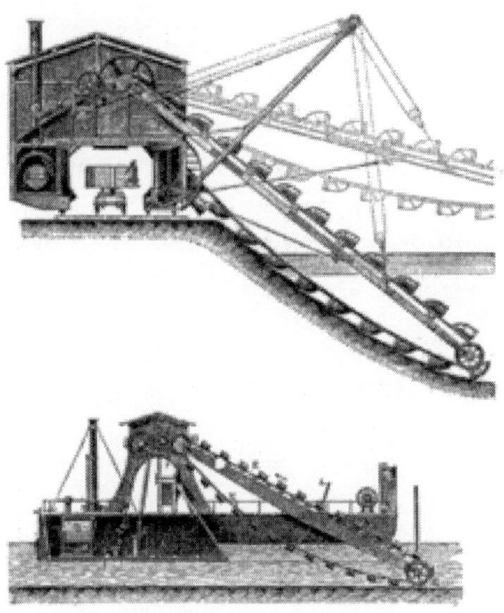

Fig. Bucket dredging

A bucket dredger is equipped with a bucket dredge, which is a device that picks up sediment by mechanical means, often with many circulating buckets attached to a wheel or chain. Some bucket dredgers and grab dredgers are powerful enough to rip out coral to make a shipping channel through coral reefs.

Clamshell

Fig. Clamshell dredging in process in Port Canaveral, Florida

A grab dredger picks up seabed material with a clam shell bucket, which hangs from an onboard crane or a crane barge, or is carried by a hydraulic arm, or is mounted like on a dragline. This technique is often used in excavation of bay mud. Most of these dredges are crane barges with spuds.

Backhoe/dipper

A backhoe/dipper dredge has a backhoe like on some excavators. A crude but usable backhoe dredger can be made by mounting a land-type backhoe excavator on a pontoon. The six largest backhoe dredgers in the world are currently the Vitruvius, the Mimar Sinan, Postnik Jakovlev (Jan De Nul), the Samson (DEME), the Simson and the Goliath (Van Oord).[citation needed] They featuredbarge-mounted excavators.

Small backhoe dredgers can be track-mounted and work from the bank of ditches. A backhoe dredger is equipped with a half-open shell. The shell is filled moving towards the machine. Usually dredges material is loaded in barges. This machine is mainly used in harbors and other shallow water.

Water injection

A water injection dredger uses a small jet to inject water under low pressure (to prevent the sediment from exploding into the surrounding waters) into the seabed to bring the sediment in suspension, which then becomes a turbidity current, which flows away down slope, is moved by a second burst of water from the WID or is carried away in natural currents. Water injection results in a lot of sediment in the water which makes measurement with most hydrographic equipment (for instance: singlebeam echosounders) difficult.

Pneumatic

These dredgers use a chamber with inlets, out of which the water is pumped with the inlets closed. It is usually suspended from a crane on land or from a small pontoon or barge. Its effectiveness depends on depth pressure.

Bed leveler

Fig. Steam dredger Bertha, built 1844, on a demonstration run in 1982

This is a bar or blade which is pulled over the seabed behind any suitable ship or boat. It has an effect similar to that of a bulldozeron land. The chain-operated steam dredger Bertha, built in 1844 to a design by Brunel and now the oldest operational steam vessel in Britain, was of this type.

Krabbelaar

This is an early type of dredger which was formerly used in shallow water in the Netherlands. It was a flat-bottomed boat with spikes sticking out of its bottom. As tide current pulled the boat, the spikes scraped seabed material loose, and the tide current washed the material away, hopefully to deeper water. Krabbelaar is Dutch for "scratcher".

Snagboat

A snagboat is designed to remove big debris such as dead trees and parts of trees from rivers and canals.

Amphibious

Some of these are any of the above types of dredger, which can operate normally, or by extending legs, also known as spuds, so it stands on the seabed with its hull out of the water. Some forms can go on land. Some of these are land-type backhoe excavators whose wheels are on long hinged legs so it can drive into shallow water and keep its cab out of water. Some of these may not have a floatable hull and, if so, cannot work in deep water. Oliver Evans (1755–1819) in 1804 invented an amphibious dredger which was America's first steam-powered road vehicle.

Submersible

These are usually used to recover useful materials from the seabed. Many of them travel on continuous track. A unique variant is intended to walk on legs on the seabed.

Fishing

Fishing dredges are used to collect various species of clams scallops, oysters or crabs from the seabed. These dredges have the form of a scoop made of chain mesh, and are towed by a fishing boat. Careless dredging can be destructive to the seabed. Nowadays some scallop dredging is replaced by collecting via scuba diving.

Police Drag[edit]

In some police departments a small dredge (sometimes called a drag) is used to find and recover objects and bodies from underwater. The bodies may be murder victims, or people who committed suicide by drowning, or victims of accidents. It is sometimes pulled by people walking on the bank. Search and rescue units also often use this type of dredge in searching for bodies of missing persons.

OTHERS TYPES OF DREDGERS

Suction Dredger

Dredging technique consist of cleaning, clearing, removal, demolition and excavation of material from the bottom of rivers, lakes, seas, wharfs and bays. Suction dredgers are widely used to removes sludge, sand, oil, and so on. Dredging, by result, can be applied on dredging's works of contaminated material or not, deepening, widening or expansion of port areas, lakes, canals, basins evolution and anchoring, mooring berths, etc.

Fig. Suction Dredger

DISPOSAL OF MATERIALS

In a "hopper dredger", the dredged materials end up in a large onboard hold called a "hopper." A suction hopper dredger is usually used for maintenance dredging. A hopper dredge usually has doors in its bottom to empty the dredged materials, but some dredges empty their hoppers by splitting the two halves of their hulls on giant hydraulic hinges. Either way, as the vessel dredges, excess water in the dredged materials is spilled off as the heavier solids settle to the bottom of the hopper. This excess water is returned to the sea to reduce weight and increase the amount of solid material (or slurry) that can be carried in one load. When the hopper is filled with slurry, the dredger stops dredging and goes to a dump site and empties its hopper.

Some hopper dredges are designed so they can also be emptied from above using pumps if dump sites are unavailable or if the dredge material is contaminated. Sometimes the slurry of dredgings and water is pumped straight into pipes which deposit it on nearby land. Other times, it is pumped into barges (also called scows), which deposit it elsewhere while the dredge continues its work.

A number of vessels, notably in the UK and NW Europe de-water the hopper to dry the cargo to enable it to be discharged onto a quayside 'dry'. This is achieved principally using self discharge bucket wheel, drag scraper or excavator via conveyor systems. When contaminated (toxic) sediments are to be removed, or large volume inland disposal sites are unavailable, dredge slurries are reduced to dry solids via a process known as dewatering. Current dewatering techniques employ either centrifuges, Geotube containers, large textile based filters or polymer flocculant/congealant based apparatus. In many projects, slurry dewatering is performed in large inland settling pits, although this is becoming less and less common as mechanical dewatering techniques continue to improve.

Similarly, many groups (most notable in east Asia) are performing research towards utilizing dewatered sediments for the production of concretes and construction block, although the high organic content (in many cases) of this material is a hindrance toward such ends.

ENVIRONMENTAL IMPACTS

Dredging can create disturbance to aquatic ecosystems, often with adverse impacts. In addition, dredge spoils may contain toxic chemicals that may have an adverse effect on the disposal area; furthermore, the process of dredging often dislodges chemicals residing in benthic substrates and injects them into the water column.

The activity of dredging can create the following principal impacts to the environment: Release of toxic chemicals (including heavy metals and PCB) from bottom sediments into the water column. Short term increases in turbidity, which can affect aquatic species metabolism and interfere with spawning. Secondary effects from water column contamination of uptake of heavy metals, DDT and other persistent organic toxins, via food chain uptake and subsequent concentrations of these toxins in higher organisms including humans.

Secondary impacts to marsh productivity from sedimentation. Tertiary impacts to avifauna which may prey upon contaminated aquatic organisms Secondary impacts to aquatic and benthic organisms' metabolism and mortality. Possible contamination of dredge spoils sites. Changes to the topography by the creation of "spoil islands" from the accumulated spoil

The nature of dredging operations and possible environmental impacts cause the industry to be closely regulated and a requirement for comprehensive regional environmental impact assessments with continuous monitoring. The U.S. Clean Water Act requires that any discharge of dredged or fill materials into "waters of the United States," including wetlands, is forbidden unless authorized by a permit issued by the Army Corps of Engineers. As a result of the potential impacts to the environment, dredging is restricted to licenced areas only with vessel activity monitored closely using automatic GPS systems.

DYNAMITE

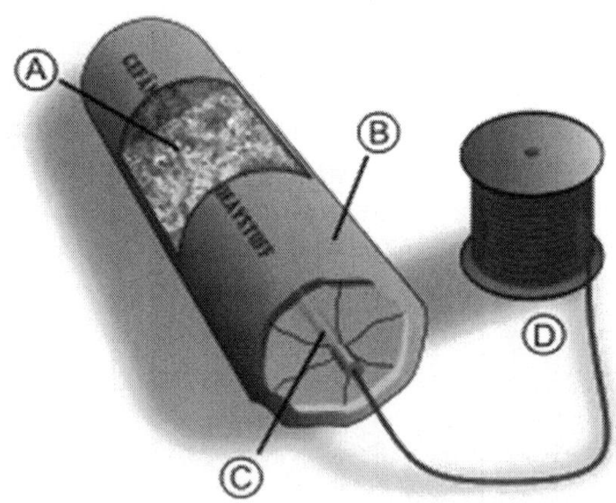

Fig. Diagram of dynamite.

(A)Sawdust (or any other type of absorbent material) soaked innitroglycerin.

(B) Protective coating surrounding the explosive material.

(C) Blasting cap.

(D) Electrical cable (or fuse) connected to the blasting cap.

Fig. Preparation of dynamite during the construction of the Douglas Dam, 1942.

Dynamite is an explosive material based on nitroglycerin, using diatomaceous earth (AmE: kieselgur; BrE: kieselguhr), or another absorbent substance such as powdered shells, clay, sawdust, or wood pulp. Dynamites using organic materials such as sawdust are less stable and such use has been generally discontinued. Dynamite was invented by the Swedish chemist and engineer Alfred Nobel in Geesthacht, Germany, and patented in 1867. Its name was coined by Nobel from the Ancient Greek word d??aμ??dýnamis, meaning "power".

Dynamite is a high explosive, which means its power comes from detonation rather than deflagration. Dynamite is mainly used in the mining, quarrying, construction, and demolition industries, and it has had some historical usage in warfare. However the unstable nature of nitroglycerin, especially if subjected to freezing, has rendered it obsolete for military uses. Public familiarity with dynamite led to metaphoric uses, such as saying that a particular issue "is political dynamite".

DESCRIPTION

Classic dynamite consists of three parts nitroglycerin, one part diatomaceous earth and a small admixture of sodium carbonate. This mixture is formed into short sticks and wrapped in paper. Nitroglycerin by itself is a very strong explosive, and in its pure form it is extremely shock-sensitive (that is, physical shock can cause it to explode), and degrades over time to even more unstable forms. This makes it highly dangerous to transport or use in its pure form.

Absorbed into diatomaceous earth or sawdust, nitroglycerin is less shock-sensitive. Over time, the dynamite will "weep" or "sweat" its nitroglycerin, which can then pool in the bottom of the box or storage area. (For that reason, explosive manuals recommend the repeated turning over boxes of dynamite in storage.) Crystals will form on the outside of the sticks causing them to be even more shock, friction or temperature sensitive. This creates a very dangerous situation. While the risk of an explosion without the use of ablasting cap is minimal for fresh dynamite, old dynamite is dangerous. Dynamite is usually sold in the form of cylinders about 8 in (20 cm) long and about 1.25 in (3.2 cm) in diameter, with a weight of about 0.5 lb troy (0.186 kg). Other sizes also exist. The maximum shelf life of nitroglycerin-based dynamite is recommended as one year from the date of manufacture under good storage conditions.

Another form of dynamite consists of nitroglycerin dissolved in nitrocellulose and a small amount of ketone. This form of dynamite is similar to cordite, and is much safer than the simple mix of nitroglycerin and diatomaceous earth. Military dynamite achieves greater stability by avoiding the use of nitroglycerin and uses much more stable chemicals.

HISTORY AND MANUFACTURE

Dynamite was invented by Alfred Nobel and was the first safely manageable explosive stronger than black powder. Nobel obtainedpatents for his invention: in England on 7 May 1867 and in Sweden on 19 October 1867. He originally sold dynamite as "Nobel's Blasting Powder". After its introduction, dynamite rapidly gained wide-scale use as a safe alternative to gunpowder andnitroglycerin. Nobel tightly controlled the patents, and unlicensed duplicating companies were quickly shut down. However, a few American businessmen got around the patent by using a slightly different formula.

South Africa

For several decades beginning in the 1940s, the largest producer of dynamite in the world was the Union of South Africa. There the De Beers company established a factory in 1902 at Somerset West. The explosives factory was later operated by AECI (African Explosives and Chemical Industries). The demand for the product came mainly from the country's vast gold mines, centered on theWitwatersrand. The factory at Somerset West was in operation in 1903 and by 1907 it was already producing 340,000 cases [22 kilograms (50 pounds) each] annually. A rival factory at Modderfontein was producing another 200,000 cases per year.

Dynamite was dangerous to manufacture. There were two large explosions at the Somerset West plant during the 1960s. Some workers died, but the loss of life was limited by the modular design of the factory and its earth works, and the planting of trees that directed the blasts upward. There were several other explosions at the Modderfontein factory. After 1985, pressure from trade unions forced AECI to phase out the production of dynamite. The factory then went on to produce ammonium nitrate emulsion-based explosives that are safer to manufacture and to handle.

United States

Dynamite was manufactured by the E. I du Pont de Nemours Company until the mid-1970s. Other American dynamite makers of that time period included the Hercules Corporation, Atlas, Trojan-US Powder, Austin, and several other smaller firms. Dynamite has been mostly phased out in favor of water gel explosives, which are cheaper to manufacture, and in many ways safer to handle.

In the United States, in 1885, the chemist Russell S. Penniman invented "ammonium dynamite", a form of explosive that used ammonium nitrate as a substitute for the more costly nitroglycerin. These dynamites were marketed with the trade name "Extra". Ammonium nitrate contains 85% of the chemical energy of nitroglycerin.

DIFFERENCE FROM TNT

Though both trinitrotoluene (TNT) and dynamite are high explosives, there is little similarity between them. Dynamite is an absorbent mixture soaked in nitroglycerin then compacted into a cylindrical shape and wrapped in paper.

TNT is a chemical compound called 2,4,6-trinitrotoluene. (Military dynamite is a dynamite substitute, also formulated without nitroglycerin, containing 75% RDX, 15% TNT, 5% SAE 10 motor oil, and 5% cornstarch, but much safer to store and handle for long time than Nobel's dynamite.) A stick of dynamite (0.19 kg, 75% nitroglycerin, no other explosives in the mixture of dynamite) contains roughly 1 MJ of energy. The energy density (joules/kilogram or J/kg) of dynamite is approximately 5 MJ/kg, compared to 4.0 MJ/kg of TNT (for TNT r.e.factor=1.00; for Nobel dynamite r.e. factor=1.25;).

BLASTING MACHINE

A blasting machine is a device that supplies electrical energy to a blasting cap to trigger a main explosive charge. It is mostly used in mining and demolition. The use of the term "machine" dates from early designs that used an electrical generator operated by winding a rotary handle or pushing down a T-handle. Modern blasting machines are battery-powered and operated by key switches and push-buttons, and do not resemble the older designs.

A typical "capacitive discharge" blasting machine works by charging a capacitor from a battery, then discharging the capacitor through an external circuit, called the firing line, to fire the blasting cap. While the machine is idle, an "internal shunt" is connected across the output terminals so that any stray voltages induced in the external circuit, for example by nearby radio transmitters, are harmlessly short-circuited without triggering the blasting cap. The machines also typically include an "abort" feature to discharge the internal capacitor without firing the cap.

Fig. Two blasting machines. At rear is one where the trigger works by depressing or raising the handle; at fore, one where the handle is twisted

HOIST (MINING)

Fig. A drum hoist (cables visible) and motor

In underground mining a hoist or winder[1] is used to raise and lower conveyances within the mine shaft. Modern hoists are normally powered using electric motors, historically with direct current drives utilizing solid-state converters (thyristors), however modern large hoists utilize alternating current drives that are variable frequency controlled. There are three principal types of hoists used in mining applications:

DRUM HOIST

Fig. A drum hoist and motor

Drum hoists are the most common type of hoist used in North America, South Africa and South America. When using a drum hoist thehoisting cable is wound around the drum when the conveyance is lifted. Single-drum hoists can be used in smaller applications, however double-drum hoists easily allow the hoisting of two conveyances in balance (*i.e.* one skip being lifted while a second skip is being lowered). Drum hoists are mounted on concrete within a hoistroom, the hoisting ropes run from the drum, up to the top of theheadframe, over a sheave wheel and down where they connect to the conveyance (cage or skip).

Advantages

Drum hoists require less routine maintenance than a friction hoist, because the haulage cable is fixed to the drum, and therefore have less downtime, and the maintenance regime is less sophisticated. Drum hoists can continue to operate if the shaft bottom gets flooded and less shaft depth is required below the loading pocket, unlike friction hoists where such flooding could cover the tail ropes etc. Because drum hoists do not have tail ropes, the hoisting system is more suited to slinging beneath a conveyance.

Disadvantages

Drum hoists take up more space than a friction hoist for the same service as all of the haulage cable must be accommodated on the drum when the hoist is fully raised. Drum hoists require rapid fluctuations in power demand, which can pose a problem if power is generated on site rather than provided through the main power grid.

FRICTION HOIST

Friction (or Koepe) hoists are the most common type of hoist used in Europe, Asia and Australia. The friction hoist was invented in 1877 byFrederick Kope. Friction hoists are mounted on the ground above the mine shaft, or at the top of the headframe. Friction hoists utilize tail ropes and counterweights and do not have the haulage rope fixed to the wheel, but instead passed around it.

Fig. A headframe housing a friction hoist

The tailropes and weights offset the need for the motor to overcome the weight of the conveyance and hoisting rope, thereby reducing the required horsepower of the hoisting motor by up to 30%, with the overall power consumption remaining the same. Friction hoists, unlike drum hoists, can and normally do use multiple ropes giving them a larger payload capacity, however since they require a larger safety factor, they are impractical for very deep shafts.

Advantages

New friction hoists are less expensive than new drum hoists, and the lead time for delivery may be shorter as there is more competition for manufacturing.

Multi-rope friction hoists have a larger lift capacity than a drum hoist. A friction hoist is smaller in diameter than a drum hoist for the same service, making it easier to ship and install than a drum hoist.

Disadvantages

Balanced friction hoists are not suitable for hoisting from multiple loading pockets on different horizons within a shaft, and are generally not suitable for deep shafts. Friction hoists can not operate at normal speeds if the shaft bottom is flooded and water reaches the tail ropes.

BEAM ENGINE

A beam engine is a type of steam engine where a pivoted overhead beam is used to apply the force from a vertical piston to a vertical connecting rod. This configuration, with the engine directly driving a pump, was first used by Thomas Newcomen around 1705 to remove water from mines in Cornwall. The efficiency of the engines was improved by engineers including James Watt who added a condenser, Jonathan Hornblower and Arthur Woolf who compounded the cylinders, and William McNaught (Glasgow) who devised a method of compounding an existing engine.

Beam engines were first used to pump water out of mines or into canals, but could be used to pump water to supplement the flow for a waterwheel powering a mill. The rotative beam engine is a later design of beam engine where the connecting rod drives a flywheel, by means of a crank (or, historically, by means of a sun and planet gear). These beam engines could be used to directly power the line-shafting in a mill. They also could be used to power steam ships.

HISTORY

The first beam engines were water-powered, and used to pump water from mines. A 'preserved' example may be seen atWanlockhead, in Scotland. Beam engines were extensively used to power pumps on the English canal system when it was expanded by means of locks early in the Industrial Revolution, and also to drain water from mines in the same period, and as winding engines. The first steam-related beam engine was developed by Thomas Newcomen. This was not, strictly speaking, steam powered, as the steam introduced below the piston was condensed to create a partial vacuum thus allowing atmospheric pressure to push down the piston. It was therefore called an Atmospheric Engine.

The Newcomen atmospheric engine was adopted by many mines in Cornwalland elsewhere, but it was relatively inefficient and consumed a large quantity of fuel. The engine was improved by Smeaton butJames Watt resolved the main inefficiencies of the Newcomen engine in his Watt steam engine by the addition of a separate condenser, thus allowing the cylinder to remain hot. Technically this was still an atmospheric engine until (under subsequent patents) he enclosed the upper part of the cylinder, introducing steam to also push the piston down.

This made it a true steam engine and arguably confirms him as the inventor of the steam engine. He also patented the centrifugal governor and the parallel motion. the latter allowed the replacement of chains round an arch head and thus allowed its use as a rotative engine. His patents remained in place until the start of the 19th Century and some say that this held back development. However in reality development had been ongoing by others and at the end of the patent period there was an explosion of new ideas and improvements. Watt's beam engines were used commercially in much larger numbers and many continued to run for 100 years or more. Watt held patents on key aspects of his engine's design, but his rotative engine was equally restricted by the patent by an other of the simple crank. The beam engine went on to be considerably improved and enlarged in the tin- and copper-rich areas of south west England, which enabled the draining of the deep mines that existed there. Consequently the Cornish beam engines became world famous, as they remain among the most massive beam engines ever constructed.

ROTATIVE BEAM ENGINES

Fig. A small rotative beam engine, built in 1870 by Thomas Horn to a design byJames Watt. The crank is visible at the front, the flywheel part-hidden by the engine. (Originally installed in a waterworks in Ashford, now operational and preserved at the Bredgar and Wormshill Light Railway.)

In a rotative beam engine, the piston is mounted vertically, and the piston rod drives the beam as before. A connecting rod from the other end of the beam, rather than driving a pump rod, now drives a flywheel.

Early Watt engines used Watt's patent sun and planet gear, rather than a simple crank, as use of the latter was protected by a patent owned by James Pickard. Once the patent had expired, the simple crank was employed universally.

Marine beam engines

The first steam-powered ships used variants of the rotative beam engine. These marine steam engines – known as side-lever,grasshopper, crosshead, or 'walking beam', among others – all varied from the original land-based machines by locating the beam or beams in different positions to take up less room on board ship.

Compounding

Compounding involves two or more cylinders; low-pressure steam from the first, high-pressure, cylinder is passed to the second cylinder where it expands further and provides more drive. This is the compound effect; the waste steam from this can produce further work if it is then passed into a condenser in the normal way.

The first experiment with compounding was conducted byJonathan Hornblower, who took out a patent in 1781. His first engine was installed at Tincroft Mine, Cornwall. It had two cylinders – one 21-inch (0.53 m) diameter with 6-foot (1.8 m) stroke and one 27-inch (0.69 m) diameter with 8-foot (2.4 m) stroke – placed alongside each other at one end of the beam. The early engines showed little performance gain: the steam pressure was too low, interconnecting pipes were of small diameter and the condenser ineffective.

At this time the laws of thermodynamics were not adequately understood, particularly the concept of absolute zero. Engineers such as Arthur Woolf were trying to tackle an engineering problem with an imperfect understanding of the physics. In particular, their valve gear was cutting-in at the wrong position in the stroke, not allowing for expansive working in the cylinder. Successful Woolf compound engines were produced in 1814, for the Wheal Abraham copper mine and the Wheal Vor tin mine.

McNaught engines

William McNaught of Glasgow, not to be confused with William McNaught of Rochdale (Petrie and McNaught), patented a compound beam engine in 1845. On a beam engine of the standard Boulton & Watt design he placed a high-pressure cylinder, on the opposite side of the beam to the existing single cylinder, where the water pump was normally fitted.

This had two important effects: it massively reduced the pressure on the beam, and the connecting steam pipe, being long, acted as an expansive receiver – the element missing in the Woolf design. This modification could be made retrospectively, and engines so modified were said to be "McNaughted". The advantages of a compound engine were not significant at pressures under 60psi, but showed at over 100psi.

DRILLING RIG

A drilling rig is a machine which creates holes in the ground. Drilling rigs can be massive structures housing equipment used to drill water wells, oil wells, or natural gas extraction wells, or they can be small enough to be moved manually by one person and are called auger. They sample sub-surface mineral deposits, test rock, soil and groundwater physical properties, and also can be used to install sub-surface fabrications, such as underground utilities, instrumentation, tunnels or wells. Drilling rigs can be mobile equipment mounted on trucks, tracks or trailers, or more permanent land or marine-based structures (such as oil platforms, commonly called 'offshore oil rigs' even if they don't contain a drilling rig). The term "rig" therefore generally refers to the complex of equipment that is used to penetrate the surface of the Earth's crust.

Fig. Upright oil derrick

Fig. Drilling Rig in Operation, North Yorkshire, United Kingdom

Drilling rigs can be: Small and portable, such as those used in mineral exploration drilling, water wells and environmental investigations. Energold's frontier S-Style drills for example focus on portability, while conventional water well rigs are far heavier for long-term drilling projects

Huge, capable of drilling through thousands of metres of the Earth's crust. Large "mud pumps" circulate drilling mud (slurry) through the drill bit and up the casing annulus, for cooling and removing the "cuttings" while a well is drilled. Hoists in the rig can lift hundreds of tons of pipe. Other equipment can force acid or sand into reservoirs to facilitate extraction of the oil or natural gas; and in remote locations there can be permanent living accommodation and catering for crews (which may be more than a hundred). Marine rigs may operate many hundreds of miles or kilometres distant from the supply base with infrequent crew rotation or cycle.

PETROLEUM DRILLING INDUSTRY

Fig. Drilling the Bakken formation in the Williston Basin

Fig. Swanson Drilling drilling directional oil wells in East Texas

Oil and natural gas drilling rigs can be used not only to identify geologic reservoirs but also to create holes that allow the extraction of oil or natural gas from those reservoirs. Primarily in onshore oil and gas fields once a well has been drilled, the drilling rig will be moved off of the well and a service rig (a smaller rig) that is purpose-built for completions will be moved on to the well to get the well on line. This frees up the drilling rig to drill another hole and streamlines the operation as well as allowing for specialization of certain services, *i.e.*, completions vs. drilling. in Middle East First Oil Well drilled in Masjed Solyman in Iran.

WATER WELL DRILLING

New portable drillcat technology uses smaller portable trailer mounted rigs with shorter 10 foot (3.0 m) drill pipe. DIY users and missionary groups and even preppers use these to drill water wells as they can be operated by 1 or 2 people with a minimal skill level. The shorter drill pipe also allows a much smaller mast, which lets you have a much smaller and lighter rig which is cheaper to ship overseas and can fit in a standard 20 foot (6.1 m) shipping container. Portable trailer mounted drilling rigs have drill ratings from 300 to 800 feet (91 to 240 m) depending on mud pump flow and pressure ratings and drill pipe sizes.

Fig. Rig drilling popularly known as 'Borewell' in operation for Water well drilling, in India.

Other, heavier, truck rigs are more complicated, thus requiring more skill to run. They're also more difficult to handle safely due to the longer 20 to 30 foot (6.1 to 9.1 m) drill pipe. Large truck rigs also require a much higher overhead clearance to operate. Large truck drills can use over 150 US gallons (570 L) or more of fuel per day, while the smaller Deeprock Style portable drills use a mere 5 to 20 US gallons (19 to 76 L) of fuel per day. The most desired portable rigs are USA made, because of the Steel quality. This makes smaller, more portable rigs preferable in remote or hard-to-reach places, and are more cost effective in this new era of high fuel prices.

History

Until internal combustion engines were developed in the late 19th century, the main method for drilling rock was muscle power of man or animal. The drilling of wells for the manufacture of salt began by the Song Dynasty in China. The well had a particularly small mouth, "as small as a small bowl." Archaeological evidence of the drilling tools used in deep-well dwelling are kept and displayed in the Zigong Salt Industry Museum.

According to Salt: A World History, a Qing Dynasty well, also located in Zigong, "continued down to 3,300 feet making it at the time the deepest drilled well in the world." Mechanised versions of this system persisted until about 1970, using a cam to rapidly raise and drop what, by then, was a steel cable up to 3 mm

In the 1970s, outside of the oil and gas industry, roller bits using mud circulation were replaced by the first pneumatic reciprocating piston Reverse Circulation (RC) drills, and became essentially obsolete for most shallow drilling, and are now only used in certain situations where rocks preclude other methods. RC drilling proved much faster and more efficient, and continues to improve with better metallurgy, deriving harder, more durable bits, and compressors delivering higher air pressures at higher volumes, enabling deeper and faster penetration. Diamond drilling has remained essentially unchanged since its inception.

MOBILE DRILLING RIGS

In early oil exploration, drilling rigs were semi-permanent in nature and the derricks were often built on site and left in place after the completion of the well. In more recent times drilling rigs are expensive custom-built machines that can be moved from well to well. Some light duty drilling rigs are like a mobile crane and are more usually used to drill water wells. Larger land rigs must be broken apart into sections and loads to move to a new place, a process which can often take weeks.

Small mobile drilling rigs are also used to drill or bore piles. Rigs can range from 100 ton continuous flight auger (CFA) rigs to small air powered rigs used to drill holes in quarries, etc. These rigs use the same technology and equipment as the oil drilling rigs, just on a smaller scale. The drilling mechanisms outlined below differ mechanically in terms of the machinery used, but also in terms of the method by which drill cuttings are removed from the cutting face of the drill and returned to surface.In Iran one company is biggest in Drilling industries NIDC (National Iranian Drilling Company)

DRILLING RIG CLASSIFICATION

There are many types and designs of drilling rigs, with many drilling rigs capable of switching or combining different drilling technologies as needed. Drilling rigs can be described using any of the following attributes:

By power used

- Mechanical—the rig uses torque converters, clutches, and transmissions powered by its own engines, often diesel
- Electric—the major items of machinery are driven by electric motors, usually with power generated on-site using internal combustion engines
- Hydraulic—the rig primarily uses hydraulic power
- Pneumatic—the rig is primarily powered by pressurize

By pipe used

- Cable—a cable is used to raise and drop the drill bit
- Conventional—uses metal or plastic drill pipe of varying types
- Coil tubing — uses a giant coil of tube and a downhole drilling motor

By height

(Rigs are differentiated by height based on how many connected pipe they are able to "stand" in the derrick when needing to temporarily remove the drill pipe from the hole. Typically this is done when changing a drill bit or when "logging" the well.)

- Single — can pull only single drill pipes. The presence or absence of vertical pipe racking "fingers" varies from rig to rig.
- Double — can hold a stand of pipe in the derrick consisting of two connected drill pipes, called a "double stand".
- Triple — can hold a stand of pipe in the derrick consisting of three connected drill pipes, called a "triple stand".
- Quadri — can store stand of pipe in the derrick composed of four connected drill pipes, called a "quadri stand".

By method of rotation or drilling method

No-rotation includes direct push rigs and most service rigs

- Rotary table — rotation is achieved by turning a square or hexagonal pipe (the "Kelly") at drill floor level.
- Top drive — rotation and circulation is done at the top of the drill string, on a motor that moves in a track along the derrick.
- Sonic — uses primarily vibratory energy to advance the drill string
- Hammer — uses rotation and percussive force (see Down-the-hole drill)

By position of derrick

- Conventional — derrick is vertical
- Slant — derrick is slanted at a 45 degree angle to facilitate horizontal drilling

DRILL TYPES

There are a variety of drill mechanisms which can be used to sink a borehole into the ground. Each has its advantages and disadvantages, in terms of the depth to which it can drill, the type of sample returned, the costs involved and penetration rates achieved. There are two basic types of drills: drills which produce rock chips, and drills which produce core samples.

Auger drilling

Auger drilling is done with a helical screw which is driven into the ground with rotation; the earth is lifted up the borehole by the blade of the screw. Hollow stem auger drilling is used for softer ground such as swamps where the hole will not stay open by itself for environmental drilling, geotechnical drilling, soil engineering and geochemistry reconnaissance work in exploration for mineraldeposits. Solid flight augers/bucket augers are used in harder ground construction drilling. In some cases, mine shafts are dug with auger drills. Small augers can be mounted on the back of a utility truck, with large augers used for sinking piles for bridge foundations. Auger drilling is restricted to generally soft unconsolidated material or weak weathered rock. It is cheap and fast.

Fig. Cable tool water well drilling rig inKimball, West Virginia. These slow rigs have mostly been replaced by rotary drilling rigs in the U.S.

Percussion rotary air blast drilling (RAB)

RAB drilling is used most frequently in the mineral exploration industry.

(This tool is also known as a Down-the-hole drill.) The drill uses a pneumatic reciprocating piston-driven "hammer" to energetically drive a heavy drill bit into the rock. The drill bit is hollow, solid steel and has ~20 mm thick tungsten rods protruding from the steel matrix as "buttons". The tungsten buttons are the cutting face of the bit.

The cuttings are blown up the outside of the rods and collected at surface. Air or a combination of air and foam lift the cuttings. RAB drilling is used primarily for mineral exploration, water bore drilling and blast-hole drilling in mines, as well as for other applications such as engineering, etc. RAB produces lower quality samples because the cuttings are blown up the outside of the rods and can be contaminated from contact with other rocks. RAB drilling at extreme depth, if it encounters water, may rapidly clog the outside of the hole with debris, precluding removal of drill cuttings from the hole. This can be counteracted, however, with the use of "stabilizers" also known as "reamers", which are large cylindrical pieces of steel attached to the drill string, and made to perfectly fit the size of the hole being drilled. These have sets of rollers on the side, usually with tungsten buttons, that constantly break down cuttings being pushed upwards.

The use of high-powered air compressors, which push 900-1150 cfm of air at 300-350 psi down the hole also ensures drilling of a deeper hole up to ~1250 m due to higher air pressure which pushes all rock cuttings and any water to the surface. This, of course, is all dependent on the density and weight of the rock being drilled, and on how worn the drill bit is.

Air core drilling

Air core drilling and related methods use hardened steel or tungsten blades to bore a hole into unconsolidated ground. The drill bit has three blades arranged around the bit head, which cut the unconsolidated ground. The rods are hollow and contain an inner tube which sits inside the hollow outer rod barrel.

The drill cuttings are removed by injection of compressed air into the hole via the annular area between the innertube and the drill rod. The cuttings are then blown back to surface up the inner tube where they pass through the sample separating system and are collected if needed. Drilling continues with the addition of rods to the top of the drill string. Air core drilling can occasionally produce small chunks of cored rock.

This method of drilling is used to drill the weathered regolith, as the drill rig and steel or tungsten blades cannot penetrate fresh rock. Where possible, air core drilling is preferred over RAB drilling as it provides a more representative sample. Air core drilling can achieve depths approaching 300 metres in good conditions. As the cuttings are removed inside the rods and are less prone to contamination compared to conventional drilling where the cuttings pass to the surface via outside return between the outside of the drill rod and the walls of the hole. This method is more costly and slower than RAB.

Cable tool drilling

Fig. SpeedStar cable tool drilling rig, Ballston Spa, New York

Cable tool rigs are a traditional way of drilling water wells. The majority of large diameter water supply wells, especially deep wells completed in bedrock aquifers, were completed using this drilling method. Although this drilling method has largely been supplanted in recent years by other, faster drilling techniques, it is still the most practicable drilling method for large diameter, deep bedrock wells, and in widespread use for small rural water supply wells.

The impact of the drill bit fractures the rock and in many shale rock situations increases the water flow into a well over rotary. Also known as ballistic well drilling and sometimes called "spudders", these rigs raise and drop a drill string with a heavy carbidetipped drilling bit that chisels through the rock by finely pulverizing the subsurface materials. The drill string is composed of the upper drill rods, a set of "jars" (inter-locking "sliders" that help transmit additional energy to the drill bit and assist in removing the bit if it is stuck) and the drill bit.

During the drilling process, the drill string is periodically removed from the borehole and a bailer is lowered to collect the drill cuttings (rock fragments, soil, etc.). The bailer is a bucket-like tool with a trapdoor in the base. If the borehole is dry, water is added so that the drill cuttings will flow into the bailer. When lifted, the trapdoor closes and the cuttings are then raised and removed. Since the drill string must be raised and lowered to advance the boring, the casing (larger diameter outer piping) is typically used to hold back upper soil materials and stabilize the borehole. Cable tool rigs are simpler and cheaper than similarly sized rotary rigs, although loud and very slow to operate. The world record cable tool well was drilled in New York to a depth of almost 12,000 feet (3,700 m). The common Bucyrus Erie 22 can drill down to about 1,100 feet (340 m). Since cable tool drilling does not use air to eject the drilling chips like a rotary, instead using a cable strung bailer, technically there is no limitation on depth.

Cable tool rigs now are nearly obsolete in the United States. They are mostly used in Africa or Third-World countries. Being slow, cable tool rig drilling means increased wages for drillers. In the United States drilling wages would average around US$200 per day per man, while in Africa it is only US$6 per day per man, so a slow drilling machine can still be used in undeveloped countries with depressed wages. A cable tool rig can drill 25 feet (7.6 m) to 60 feet (18 m) of hard rock a day. A newer rotary drillcat top head rig equipped with down-the-hole (DTH) hammer can drill 500 feet (150 m) or more per day, depending on size and formation hardness.

Reverse circulation (RC) drilling

Fig. Reverse Circulation (RC) rig, outsideNewman, Western Australia

Fig. Track mounted Reverse Circulation rig (side view).

RC drilling is similar to air core drilling, in that the drill cuttings are returned to surface inside the rods. The drilling mechanism is apneumatic reciprocating piston known as a "hammer" driving a tungsten-steel drill bit. RC drilling utilises much larger rigs and machinery and depths of up to 500 metres are routinely achieved. RC drilling ideally produces dry rock chips, as large aircompressors dry the rock out ahead of the advancing drill bit. RC drilling is slower and costlier but achieves better penetration than RAB or air core drilling; it is cheaper than diamond coring and is thus preferred for most mineral exploration work.

Reverse circulation is achieved by blowing air down the rods, the differential pressure creating air lift of the water and cuttings up the "inner tube", which is inside each rod. It reaches the "divertor" at the top of the hole, then moves through a sample hose which is attached to the top of the "cyclone". The drill cuttings travel around the inside of the cyclone until they fall through an opening at the bottom and are collected in a sample bag. The most commonly used RC drill bits are 5-8 inches (13–20 cm) in diameter and have round tungsten 'buttons' that protrude from the bit, which are required to drill through shale and abrasive rock.

As the buttons wear down, drilling becomes slower and the rod string can potentially become bogged in the hole. This is a problem as trying to recover the rods may take hours and in some cases weeks. The rods and drill bits themselves are very expensive, often resulting in great cost to drilling companies when equipment is lost down the bore hole. Most companies will regularly re-grind the buttons on their drill bits in order to prevent this, and to speed up progress. Usually, when something is lost (breaks off) in the hole, it is not the drill string, but rather from the bit, hammer, or stabilizer to the bottom of the drill string (bit). This is usually caused operator error, over-stressed metal, or adverse drilling conditions causing downhole equipment to get stuck in a part of the hole.

Although RC drilling is air-powered, (water is also used, to reduce dust, keep the drill bit cool, and assist in pushing cutting back upwards, but also when "collaring" a new hole. A mud called "Liqui-Pol" is mixed with water and pumped into the rod string, down the hole. This helps to bring up the sample to the surface by making the sand stick together. Occasionally, "Super-Foam" (a.k.a. "Quik-Foam") is also used, to bring all the very fine cuttings to the surface, and to clean the hole. When the drill reaches hard rock, a "collar" is put down the hole around the rods, which is normally PVC piping. Occasionally the collar may be made from metal casing. Collaring a hole is needed to stop the walls from caving in and bogging the rod string at the top of the hole. Collars may be up to 60 metres deep, depending on the ground, although if drilling through hard rock a collar may not be necessary. Reverse circulation rig setups usually consist of a support vehicle, an auxiliary vehicle, as well as the rig itself. The support vehicle, normally a truck, holds diesel and water tanks for resupplying the rig. It also holds other supplies needed for maintenance on the rig. The auxiliary is a vehicle, carrying an auxiliary engine and a booster engine. These engines are connected to the rig by high pressure air hoses. Although RC rigs have their own booster and compressor to generate air pressure, extra power is needed which usually isn't supplied by the rig due to lack of space for these large engines. Instead, the engines are mounted on the auxiliary vehicle. Compressors on an RC rig have an output of around 1000 cfm at 500 psi (500 L•s-1 at 3.4 MPa). Alternatively, stand-alone air compressors which have an output of 900-1150cfm at 300-350 psi each are used in sets of 2, 3, or 4, which are all routed to the rig through a multi-valve manifold.

Diamond core drilling

Fig. Multi-combination drilling rig (capable of both diamond and reverse circulation drilling). Rig is currently set up for diamond drilling.

Diamond core drilling (exploration diamond drilling) utilizes an annular diamond-impregnated drill bit attached to the end of hollow drill rods to cut a cylindrical core of solid rock. The diamonds used to make diamond core bits are a variety of sizes, fine to microfine industrial grade diamonds, and the ratio of diamonds to metal used in the matrix affects the performance of the bits cutting ability in different types of rock formations . The diamonds are set within a matrix of varying hardness, from brass to high-grade steel. Matrix hardness, diamond size and dosing can be varied according to the rock which must be cut. The bits made with hard steel with a low diamond count and are ideal for softer highly fractured rock while others made of softer steels and high dimond ratio are good for coring in hard solid rock. Holes within the bit allow water to be delivered to the cutting face. This provides three essential functions — lubrication, cooling, and removal of drill cuttings from the hole.

Diamond drilling is much slower than reverse circulation (RC) drilling due to the hardness of the ground being drilled. Drilling of 1200 to 1800 metres is common and at these depths, ground is mainly hard rock. Techniques vary among drill operators and what the rig they are using is capable of, some diamond rigs need to drill slowly to lengthen the life of drill bits and rods, which are very expensive and time consuming to replace at extremely deep depths. As a diamond drill rig cores deeper and deeper the time consuming part of the process is not cutting 5 to 10 more feet of rock core but the retrieval of the core with the whire line & overshot tool.

Core samples are retrieved via the use of a core tube, a hollow tube placed inside the rod string and pumped with water until it locks into the core barrel. As the core is drilled, the core barrel slides over the core as it is cut. An "overshot" attached to the end of the winch cable is lowered inside the rod string and locks on to the backend (aka head assembly), located on the top end of the core barrel. The winch is retracted, pulling the core tube to the surface. The core does not drop out of the inside of the core tube when lifted because either a split ring core lifter or basket retainer allow the core to move into, but not back out of the tube.

Fig. Diamond core drill bits

Once the core tube is removed from the hole, the core sample is then removed from the core tube and catalogued. The Driller's assistant unscrews the backend off the core tube using tube wrenches, then each part of the tube is taken and the core is shaken out into core trays. The core is washed, measured and broken into smaller pieces using a hammer or sawn through to make it fit into the sample trays. Once catalogued, the core trays are retrieved by geologists who then analyse the core and determine if the drill site is a good location to expand future mining operations. Diamond rigs can also be part of a multi-combination rig. Multi-combination rigs are a dual setup rig capable of operating in either a reverse circulation (RC) and diamond drilling role (though not at the same time). This is a common scenario where exploration drilling is being performed in a very isolated location. The rig is first set up to drill as an RC rig and once the desired metres are drilled, the rig is set up for diamond drilling. This way the deeper metres of the hole can be drilled without moving the rig and waiting for a diamond rig to set up on the pad.

Direct push rigs

Direct push technology includes several types of drilling rigs and drilling equipment which advances a drill string by pushing or hammering without rotating the drill string. While this does not meet the proper definition of drilling, it does achieve the same result — a borehole. Direct push rigs include both cone penetration testing (CPT) rigs and direct push sampling rigs such as a PowerProbe or Geoprobe. Direct push rigs typically are limited to drilling in unconsolidated soil materials and very soft rock.

CPT rigs advance specialized testing equipment (such as electronic cones), and soil samplers using large hydraulic rams. Most CPT rigs are heavily ballasted (20 metric tons is typical) as a counter force against the pushing force of the hydraulic rams which are often rated up to 20 kN. Alternatively, small, light CPT rigs and offshore CPT rigs will use anchors such as screwed-in ground anchors to create the reactive force. In ideal conditions, CPT rigs can achieve production rates of up to 250–300 meters per day.

Direct push drilling rigs use hydraulic cylinders and a hydraulic hammer in advancing a hollow core sampler to gather soil and groundwater samples. The speed and depth of penetration is largely dependent on the soil type, the size of the sampler, and the weight and power of the rig. Direct push techniques are generally limited to shallow soil sample recovery in unconsolidated soil materials. The advantage of direct push technology is that in the right soil type it can produce a large number of high quality samples quickly and cheaply, generally from 50 to 75 meters per day. Rather than hammering, direct push can also be combined with sonic (vibratory) methods to increase drill efficiency.

Hydraulic rotary drilling

Oil well drilling utilises tri-cone roller, carbide embedded, fixed-cutter diamond, or diamond-impregnated drill bits to wear away at the cutting face. This is preferred because there is no need to return intact samples to surface for assay as the objective is to reach a formation containing oil or natural gas. Sizable machinery is used, enabling depths of several kilometres to be penetrated. Rotating hollow drill pipes carry down bentonite and barite infused drilling muds to lubricate, cool, and clean the drilling bit, control downhole pressures, stabilize the wall of the borehole and remove drill cuttings. The mud travels back to the surface around the outside of the drill pipe, called the annulus. Examining rock chips extracted from the mud is known as mud logging. Another form of well logging is electronic and is frequently employed to evaluate the existence of possible oil and gas deposits in the borehole. This can take place while the well is being drilled, using Measurement While Drilling tools, or after drilling, by lowering measurement tools into the newly drilled hole.

The rotary system of drilling was in general use in Texas in the early 1900s. It is a modification of one invented by Fauvelle in 1845, and used in the early years of the oil industry in some of the oil-producing countries in Europe. Originally pressurized water was used instead of mud, and was almost useless in hard rock before the diamond cutting bit. The main breakthrough for rotary drilling came in 1901, when Anthony Francis Lucas combined the use of a steam-driven rig and of mud instead of water in theSpindletop discovery well.

The drilling and production of oil and gas can pose a safety risk and a hazard to the environment from the ignition of the entrained gas causing dangerous fires and also from the risk of oil leakage polluting water, land and groundwater. For these reasons, redundant safety systems and highly trained personnel are required by law in all countries with significant production.

Sonic (vibratory) drilling

A sonic drill head works by sending high frequency resonant vibrations down the drill string to the drill bit, while the operator controls these frequencies to suit the specific conditions of the soil/rock geology. Vibrations may also be generated within the drill head. The frequency is generally between 50 and 120 hertz (cycles per second) and can be varied by the operator. Resonance magnifies the amplitude of the drill bit, which fluidizes the soil particles at the bit face, allowing for fast and easy penetration through most geological formations. An internal spring system isolates these vibrational forces from the rest of the drill rig.

LIMITS OF THE TECHNOLOGY

Drill technology has advanced steadily since the 19th century. However, there are several basic limiting factors which will determine the depth to which a bore hole can be sunk. All holes must maintain outer diameter; the diameter of the hole must remain wider than the diameter of the rods or the rods cannot turn in the hole and progress cannot continue.

Friction caused by the drilling operation will tend to reduce the outside diameter of the drill bit. This applies to all drilling methods, except that in diamond core drilling the use of thinner rods and casing may permit the hole to continue. Casing is simply a hollow sheath which protects the hole against collapse during drilling, and is made of metal or PVC. Often diamond holes will start off at a large diameter and when outside diameter is lost, thinner rods put down inside casing to continue, until finally the hole becomes too narrow.

Alternatively, the hole can be reamed; this is the usual practice in oil well drilling where the hole size is maintained down to the next casing point. For percussion techniques, the main limitation is air pressure. Air must be delivered to the piston at sufficient pressure to activate the reciprocating action, and in turn drive the head into the rock with sufficient strength to fracture and pulverise it. With depth, volume is added to the in-rod string, requiring larger compressors to achieve operational pressures. Secondly, groundwater is ubiquitous, and increases in pressure with depth in the ground. The air inside the rod string must be pressurised enough to overcome this water pressure at the bit face.

Then, the air must be able to carry the rock fragments to surface. This is why depths in excess of 500 m for reverse circulation drilling are rarely achieved, because the cost is prohibitive and approaches the threshold at which diamond core drilling is more economic. Diamond drilling can routinely achieve depths in excess of 1200 m. In cases where money is no issue, extreme depths have been achieved, because there is no requirement to overcome water pressure. However, water circulation must be maintained to return the drill cuttings to surface, and more importantly to maintain cooling and lubrication of the cutting surface of the bit; while at the same time reduce friction on the steel walls of the rods turning against the rock walls of the hole. When water return is lost the rods will vibrate, this is called "rod chatter", and that will damage the drill rods, crack the joints.

Without sufficient lubrication and cooling, the matrix of the drill bit will soften. While diamond is the hardest substance known, at 10 on the Mohs hardness scale, it must remain firmly in the matrix to achieve cutting. Weight on bit, the force exerted on the cutting face of the bit by the drill rods in the hole above the bit, must also be monitored. A unique drilling operation in deep ocean water was named Project Mohole.

RESEARCH OF NEW DRILLING TECHNOLOGIES

Research includes technologies based on the utilization of water jet, chemical plasma, hydrothermal spallation or laser. Utilization of high energetic electrical plasma shows promise in deep drilling applications. This approach has potential to replace conventional drilling technologies. It would be able to produce boreholes with large constant diameter without frequent replacement of the drill bits.

CAUSES OF DEVIATION

Most drill holes deviate slightly from their planned trajectory. This is because of the torque of the turning bit working against the cutting face, because of the flexibility of the steel rods and especially the screw joints, because of reaction to foliation and structure within the rock, and because of refraction as the bit moves into different rock layers of varying resistance. Additionally, inclined holes will tend to deviate upwards because the drill rods will lie against the bottom of the bore, causing the drill bit to be slightly inclined from true. It is because of deviation that drill holes must be surveyed if deviation will impact the usefulness of the information returned. Sometimes the surface location can be offset laterally to take advantage of the expected deviation tendency, so the bottom of the hole will end up near the desired location. Oil well drilling commonly uses a process of controlled deviation called directional drilling (*e.g.*, when several wells are drilled from one surface location).

Rig Equipment

Drilling rigs typically include at least some of the following items: See Drilling rig (petroleum) for a more detailed description.

Blowout Preventers: (BOPs)

The equipment associated with a rig is to some extent dependent on the type of rig but (#23 & #24) are devices installed at the wellhead to prevent fluids and gases from unintentionally escaping from the borehole. #23 is the annular (often referred to as the "Hydril", which is one manufacturer) and #24 is the pipe rams and blind rams. In the place of #24 Variable bore rams or VBRs can be used. These offer the same pressure and sealing capacity found in standard pipe rams, while offering the versatility of sealing on various sizes of drill pipe, production tubing and casing without changing standard pipe rams. Normally VBRs are used when utilizing a tapered drill string (when different size drill pipe is used in the complete drill string).

- *Centrifuge*: an industrial version of the device that separates fine silt and sand from the drilling fluid.
- *Solids control*: solids control equipments for preparing drilling mud for the drilling rig.
- *Chain tongs*: wrench with a section of chain, that wraps around whatever is being tightened or loosened. Similar to a pipe wrench.
- *Degasser*: a device that separates air and/or gas from the drilling fluid.
- *Desander/desilter*: contains a set of hydrocyclones that separate sand and silt from the drilling fluid.
- *Drawworks*: (#7) is the mechanical section that contains the spool, whose main function is to reel in/out the drill line to raise/lower the traveling block (#11).
- *Drill Bit*: (#26) is a device attached to the end of the drill string that

breaks apart the rock being drilled. It contains jets through which the drilling fluid exits.

- Drill pipe: (#16) joints of hollow tubing used to connect the surface equipment to the bottom hole assembly (BHA) and acts as a conduit for the drilling fluid. In the diagram, these are "stands" of drill pipe which are 2 or 3 joints of drill pipe connected together and "stood" in the derrick vertically, usually to save time while tripping pipe.
- Elevators: a gripping device that is used to latch to the drill pipe or casing to facilitate the lowering or lifting (of pipe or casing) into or out of the borehole.
- Mud motor: a hydraulically powered device positioned just above the drill bit used to spin the bit independently from the rest of the drill string.
- Mud pump: (#4) reciprocal type of pump used to circulate drilling fluid through the system.
- Mud tanks: (#1) often called mud pits, provides a reserve store of drilling fluid until it is required down the wellbore.
- Rotary table: (#20) rotates the drill string along with the attached tools and bit.
- Shale shaker: (#2) separates drill cuttings from the drilling fluid before it is pumped back down the borehole.

WELLHEAD

A wellhead is the component at the surface of an oil or gas well that provides the structural and pressure-containing interface for the drilling and production equipment.

Fig. Wellhead gas storage, Etzel Germany

The primary purpose of a wellhead is to provide the suspension point and pressure seals for the casing strings that run from the bottom of the hole sections to the surface pressure control equipment. While drilling the oil well, surface pressure control is provided by a blowout preventer (BOP). If the pressure is not contained during drilling operations by the column of drilling fluid, casings, wellhead, and BOP, a well blowout could occur.

Once the well has been drilled, it is completed to provide an interface with the reservoir rock and a tubular conduit for the well fluids. The surface pressure control is provided by a Christmas tree, which is installed on top of the wellhead, with isolation valvesand choke equipment to control the flow of well fluids during production.

Wellheads are typically welded onto the first string of casing, which has been cemented in place during drilling operations, to form an integral structure of the well. In exploration wells that are later abandoned, the wellhead may be recovered for refurbishment and re-use. Offshore, where a wellhead is located on the production platform it is called a surface wellhead, and if located beneath the water then it is referred to as a subsea wellhead or mudline wellhead.

FUNCTIONS

A wellhead serves numerous functions, some of which are:

1. Provide a means of casing suspension. (Casing is the permanently installed pipe used to line the well hole for pressure containment and collapse prevention during the drilling phase).
2. Provides a means of tubing suspension. (Tubing is removable pipe installed in the well through which well fluids pass).
3. Provides a means of pressure sealing and isolation between casing at surface when many casing strings are used.
4. Provides pressure monitoring and pumping access to annuli between the different casing/tubing strings.
5. Provides a means of attaching a blowout preventer during drilling.
6. Provides a means of attaching a Christmas tree for production operations.
7. Provides a reliable means of well access.
8. Provides a means of attaching a well pump,

COMPONENTS

The primary components of a wellhead system are:

- Casing head
- Casing spools
- Casing hangers
- Packoffs (isolation) seals
- Bowl protectors / wear bushings
- Test plugs
- Mudline suspension systems

- Tubing heads
- Tubing hangers
- Tubing head adapters

Design specification

The oil industry specifications for wellhead systems (materials, dimensions, test procedures and pressure ratings etc.) are:

1. API 6A Specification for Wellhead and Christmas Tree Equipment
2. ISO 10423 Wellhead and Christmas Tree Equipment

In general well heads are five nominal ratings of wellheads: 2, 3, 5, 10 and 15 (x1000) psi working pressure. They have an operating temperature range of -50 to +250 degrees Fahrenheit. They are used in conjunction with ring type seal gaskets. In general the yield strength of the materials range from 36000 to 75000 psi.

5

Mining Processing

Material extracted from the earth is thoroughly processed in the mining process. Mining products used on a daily basis, such as metals, are the end result of a long series of transformations. The series of transformations starts with the rock in the earth's crust. Various minerals are formed in the earth's crust as the result of different deposits and compound combinations. These mineral deposits are called ore. Most ore contains unnecessary minerals that are removed in the mineral process. The type of mining process used depends on the type of mineral being processed. Minerals are usually categorized as rock materials, industrial minerals (*e.g.* clay), base metals (*e.g.* copper) and precious metals (*e.g.* gold). Other materials used by industry, such as coal, potash, iron ore and diamonds, are also mined.

BEGINNING OF THE PROCESS

Rock material is mined using explosives and quarrying. Large boulders are transported to big crushers that crush the rock material into smaller grain size. Various sizes of crushers, both stationary and mobile, are used at mines. Among the types used are primary gyratory crushers, impact crushers, gyratory crushers, cone crushers and jaw crushers. The material to be crushed is fed into the crusher either via a conveyor, an excavator or a wheeled loader (reclaimer) After crushing, the rock material is ground in grinding mills. There are a variety of different grinding mills, including rod and ball mills, semiautogenous (SAG) mills and autogenous mills. In the grinding mills, the impact of the minerals grinding against each other and against the steel balls in the mill breaks down the size of the minerals even further. The mining process also includes a vibrating screen that sorts the crushed material into desired grain sizes. Rock material is extremely hard and thus it wears down the crushers and grinders. Therefore the wear parts, like the various blades and rotors and the liners of grinding mills, must be replaced regularly.

Mineral process

In the mineral process valuable minerals are concentrated by removing unnecessary substances from the excavated rock material. The first phase of the process involves reducing the size of the mineral. This is done in crushers and grinding mills where the ore is broken down into smaller pieces. When the fragments of ore are small enough, the pieces containing the most valuable minerals are separated from those containing mostly unneeded minerals. The separation process involves leaching, emission, gravity methods and magnetic separation.

Emission chambers and pipes, coils and magnetic separation equipment are used in the separation process. Water must also be extracted from the concentrated product. Coils, filters, thickeners, sedimentation basins and dryers are used to do this.

Metal extraction

In the next phase, metal is extracted from the mineral. Extraction methods vary depending on the product, but the main methods are pyrometallurgy (a metal production method employing high temperatures), hydrometallurgy (producing metal by leaching the raw material and flocculating the pure metal from the solution) and electrometallurgy (a metal production method applying electricity). Electrolysis, a forced oxidation-reduction reaction achieved using electric current, is an integral part of electrometallurgy. Electrolysis can be used to produce very pure elements.

Metal refining

After the extraction phase, the metals still contain impurities, so they must be purified. The refining process uses heat, chemicals and electricity. Electrowinning is used in gold refining.

MINING – OPEN PIT AND UNDERGROUND

To define the ore from the waste rock, samples are taken and assayed. Assay results are used to mark out areas of ore and waste rock, which are mined separately. Some of the harder areas require blasting to loosen the rock prior to excavation by hydraulic diggers. Dump trucks haul the rock to the primary crushers in the open pit or to the ore stockpiles from underground.

Crushing

The primary crushers located at the open pit mine site, receive ore and waste at separate times. They break the larger rocks down to a size suitable for transport on the conveyor belt.

Transport

A rubber belted conveyor transports the ore and waste rock from the open pit, approximately two kilometres, via a tunnel through Union Hill, to the mill and waste rock embankments. Large electromagnets remove any steel debris excavated from the old Martha Mine workings.

Grinding and sizing

Ore from both open pit and underground is stockpiled separately at the mill before being fed into a S.A.G. mill with lime, water and steel balls. The larger particles from this mill are returned to the S.A.G. mill for more grinding. The finer particles receive more grinding in a ball mill, and are size classified to give a final product of 80% <70 microns.

Leaching and adsorption

A slurry of ground ore, water and a weak cyanide solution is fed into large steel leach tanks where the gold and silver are dissolved. Following this leaching process the slurry passes through six adsorption tanks containing carbon granules which adsorb the gold and silver. This process removes 93% of the gold and 70% of the silver.

Elution and electrowinning

The loaded carbon is fed into an elution column where the bullion is washed off. The barren carbon is recycled. The wash solution—pregnant electrolyte—is passed through electrowinning cells where gold and silver is won onto stainless steel cathodes.

Bullion production

The loaded cathodes are rinsed to yield a gold and silver bearing sludge which is dried, mixed with fluxes and put into the furnace. After several hours the molten material is poured into a cascade of moulds producing bars of doré bullion.

Water Treatment

Some water from dewatering the mine, from the embankment underdrains and decant water from the tailings pond is recycled for use in the grinding circuit. Excess water is pumped to the water treatment plant and treated to the required standards before discharge into the Ohinemuri River.

Tailings Disposal

Waste rock from the open pit mine is used to build the embankment structures. The embankments retain the tailings slurry in a pond where solids settle and compact. Water is decanted off and used in the process plant or treated before it is discharged.

MINERAL PROCESSING

In the field of extractive metallurgy, mineral engineering,mineral processing, also known as mineral dressing or ore dressing, is the process of separating commercially valuable minerals from their ores.

HISTORY

Fig. A set of Cornish stamps

Before the advent of heavy machinery the raw ore was broken up using hammers wielded by hand, a process called "spalling". Before long, mechanical means were found to achieve this. For instance, stamp mills were used in Samarkand as early as 973. They were also in use in medieval Persia. By the 11th century, stamp mills were in widespread use throughout the medieval Islamic world, from Islamic Spain and North Africa in the west to Central Asia in the east. A later example was the Cornish stamps, consisting of a series of iron hammers mounted in a vertical frame, raised by cams on the shaft of a waterwheel and falling on to the ore under gravity.

The simplest method of separating ore from gangue consists of picking out the individual crystals of each. This is a very tedious process, particularly when the individual particles are small. Another comparatively simple method relies on the various minerals having different densities, causing them to collect in different places: metallic minerals (being heavier) will drop out of suspension more quickly than lighter ones, which will be carried further by a stream of water.

The process of panning and sifting for gold uses both of these methods. Various devices known as 'buddles' were used to take advantage of this property.[when?] Later, more advanced machines were used such as the Frue vanner, invented in 1874. Other equipment used historically includes the hutch, a trough used with some ore-dressing machines and the keeve or kieve, a large tub used for differential settlement.

UNIT OPERATIONS

Mineral processing can involve four general types of unit operation: comminution – particle size reduction; sizing – separation of particle sizes by screening or classification;concentration by taking advantage of physical and surface chemical properties; and dewatering – solid/liquid separation. In all of these processes, the most important considerations are the economics of the processes and this is dictated by the grade and recovery of the final product. To do this, the mineralogy of the ore needs to be considered as this dictates the amount of liberation required and the processes that can occur. The smaller the particles processes, the greater the theoretical grade and recovery of the final product, but this however is difficult to do with fine particles as they prevent certain concentration processes from occurring.

Comminution

Comminution is particle size reduction of materials. Comminution may be carried out on either dry materials or slurries. Crushing and grinding are the two primary comminution processes. Crushing is normally carried out on "run-of-mine" ore, while grinding (normally carried out after crushing) may be conducted on dry or slurried material.

Sizing

Fig. Sizer 2000 for screening coarse to small particles

Sizing is the general term for separation of particles according to their size. The simplest sizing process is screening, or passing the particles to be sized through a screen or number of screens. Screening equipment can include grizzlies, bar screens,wedge wire screens, radial sieves, banna screens, multi-deck screens, vibratory screen, fine screens, flip flop screens and wire mesh screens.

Screens can be static (typically the case for very coarse material), or they can incorporate mechanisms to shake or vibrate the screen. Some considerations in this process includes the screen material, the aperture size, shape and orientation, the amount of near sized particles, the addition of water, the amplitude and frequency of the vibrations, the angle of inclination, the presence of harmful materials, like steel and wood, and the size distribution of the particles.

Classification refers to sizing operations that exploit the differences in settling velocities exhibited by particles of different size. Classification equipment may include ore sorters, gas cyclones, hydrocyclones, rotating trommels, rake classifiers or fluidized classifiers.

An important factor in both comminution and sizing operations is the determination of the particle size distribution of the materials being processed, commonly referred to as particle size analysis. Many techniques for analyzing particle size are used, and the techniques include both off-line analyses which require that a sample of the material be taken for analysis and on-line techniques that allow for analysis of the material as it flows through the process.

Concentration

There are a number of ways to increase the concentration of the wanted minerals: in any particular case the method chosen will depend on the relative physical and surface chemical properties of the mineral and the gangue. Concentration is defined as the number of moles of a solute in a volume of the solution.

Gravity concentration

Historically the earliest method used,particles can be classified based on their specific gravity. Air is the main fluid medium used for the process. Gravity concentration processes include:

- Heavy media or dense media separation (these include, baths, drums, larcodems, dyana whirlpool separators, and dense medium cyclones)
- Shaking tables, such as the Wilfley table
- Spiral separators
- Reflux Classifier
- Batac jigs
- Centrifugal bowl concentrators, such as the Knelson concentrator and Falcon Concentrator
- Jig concentrators are continuous processing gravity concentration devices using a pulsating fluidized bed.(RMS-Ross Corp. Circular Jig Plants)
- Multi gravity separators (Falcon Concentrator, Knelson, Mozley (Conseptor Enhanced Gravity Separator) and the Kelsey Jig)
- Inline pressure Jigs
- Reichert Cones

These processes can be classified as either dense medium separation or gravity separation. The difference between the two that gravity separation does not use a dense medium to operate, only water or air. Dense medium separation can be performed with a variety of mediums.

These include, organic liquids, aqueous solutions, suspensions in water and suspensions in air. Of these, most industrial processes use suspensions in water. The organic liquids are not used due to their toxicity and difficulties in handling. The aqueous solution as a dense medium is used in coal processing in the form of a belknap wash and the suspension in air is used in water-deficient areas, like china, where sand is used to separate coal from the gangue minerals. The dense medium separation is also classified as absolute gravity separation as the sinks and the floats travel in different directions. The gravity separation is also called relative gravity separation as they separate particles due to their differences in the magnitude of the particle response to a driving force.

These processes can also be classified into multi-G and single G processes. The difference is the magnitude of the driving force for the separation. Multi-G processes allow the separation of fine particles to occur and these particles can be in the range of 10 to 50 micron. The single G process are only capable of processing particles that are greater than 80 micron in diameter. Of the gravity separation processes, the spiral concentrators and circular jigs are two of the most economical due to their simplicity and use of space. They operate by flowing film separation and can either use washwater or be washwater-less. The washwater spirals separate particles more easily but can have issues with entrainment of gangue with the concentrate produced..

Fig. Froth flotation cells used to concentrate copper and nickel sulfide minerals, Falconbridge, Ontario.

Froth flotation

Froth flotation is an important concentration process. This process can be used to separate any two different particles and operated by the surface chemistry of the particles. In flotation, bubbles are introduced into a pulp and the bubbles rise through the pulp. In the process, hydrophobic particles become bound to the surface of the bubbles. The driving force for this attachment is the change in the surface free energy when the attachment occurs. These bubbles rise through the slurry and are collected from the surface. To enable these particles to attach, careful consideration of the chemistry of the pulp needs to be made. These considerations include the pH, Eh and the presence of flotation reagents. The pH is important as it changes the charge of the particles surface and the Eh affects the chemisorption of collectors on the surface of the particles.

The addition of flotation reagents also affects the operation of these processes. The most important chemical that is added is the collector, This chemical binds to the surface of the particles as it is a surfactant. The main considerations in this chemical is the nature of the head group and the size of the hydrocarbon chain. The hydrocarbon tail needs to be short to maximize the selectivity of the desired mineral and the headgroup dictates which minerals it attaches to. The frothers are another important chemical addition to the pulp at it enables stable bubbles to be formed. This is important as if the bubble coalesce, minerals fall off their surface. The bubbles however should not be too stable as this prevents easy transportation and dewatering of the concentrate formed. The mechanism of these frothers is not completely known and further research into their mechanisms is being performed.

Depressants and activators are used to selectively separate one mineral from another. Depressants inhibit the flotation of one mineral or minerals while activators enable the flotation of others. Examples of these include CN-, used to depress all sulfides but galena and this depressant is believed to operate by changing the solubility of chemisorbed and physisorbed collectors on sulfides. This theory originates from Russia. An example of an activator is Cu2+ ions, used for the flotation of sphalerite. There are a number of cells able to be used for the flotation of minerals. these include flotation columns and mechanical flotation cells. The flotation columns are used for finer minerals and they typically have a higher grade and lower recovery of minerals than mechanical flotation cells. The cells in use at the moment can exceed 300 m3. This is done as they are cheaper per unit volume than smaller cells, but they are not able to be controlled as easily as smaller cells.

This process was invented in the 19th century in Australia. It was used to recover a sphalerite concentrate from tailings, produced using gravity concentration. Further improvements have come from Australia in the form of the Jameson Cell, developed at the University of Newcastle, Australia. This operated by the use of a plunging jet that generates fine bubbles. These fine bubbles have a higher kinetic energy and as such they can be used for the flotation of fine grained minerals, such as those produced by the isamill.

Electrostatic separation

There are two main types of electrostatic separators. These work in similar ways, but the forces applied to the particles are different and these forces are gravity and electrostatic attraction. The two types are electrodynamic separators (or high tension rollers) or electrostatic separators. In high tension rollers, particles are charged by a corona discharge. This charges the particles that subsequently travel on a drum. The conducting particles lose their charge to the drum and are removed from the drum with centripetal acceleration. Electrostatic plate separators work by passing a stream of particles past a charged anode. The conductors lose electrons to the plate and are pulled away from the other particles due to the induced attraction to the anode. These separators are used for particles between 75 and 250 micron and for efficient separation to occur, the particles need to be dry, have a close size distribution and uniform in shape. Of these considerations, one of the most important is the water content of the particles. This is important as a layer of moisture on the particles will render the non-conductors as conductors as the layer of the water is conductive.

Electrostatic plate separators are usually used for streams that have small conductors and coarse non-conductors. The high tension rollers are usually used for streams that have coarse conductors and fine non-conductors. These separators are commonly used for separating mineral sands, an example of one of these mineral processing plants is the CRL processing plant at Pinkenba in Brisbane Queensland. In this plant, zircon, rutile and ilmenite are separated from the silica gangue. In this plant, the separation is performed in a number of stages with roughers, cleaners, scavengers and recleaners.

Magnetic separation

Magnetic separation is a process in which magnetically susceptible material is extracted from a mixture using a magnetic force. This separation technique can be useful in mining iron as it is attracted to a magnet. In mines where wolframite was mixed with cassiterite, such as South Crofty and East Pool mine in Cornwall or with bismuth such as at the Shepherd and Murphy mine in Moina, Tasmania, magnetic separation was used to separate the ores. At these mines a device called a Wetherill's Magnetic Separator (invented by John Price Wetherill, 1844–1906) was used. In this machine the raw ore, after calcination was fed onto a moving belt which passed underneath two pairs of electromagnets under which further belts ran at right angles to the feed belt. The first pair of electromagnets was weakly magnetised and served to draw off any iron ore present. The second pair were strongly magnetised and attracted the wolframite, which is weakly magnetic. These machines were capable of treating 10 tons of ore a day.This process of separating magnetic substances from the non-magnetic substances in a mixture with the help of a magnet is called magnetic separation..

This process operates by moving particles in a magnetic field. The force experienced in the magnetic field is given by the equation f=m/k.H.dh/dx. with k=magnetic susceptibility, H-magnetic field strength, and dh/dx being the magnetic field gradient. As seen in this equation, the separation can be driven in two ways, either through a gradient in a magnetic field or the strength of a magnetic field. The different driving forces are used in the different concentrators. These can be either with water or without. Like the spirals, washwater aids in the separation of the particles while increases the entrainment of the gangue in the concentrate.

Automated Ore Sorting

Modern, automated sorting applies optical sensors (visible spectrum, near infrared, X-ray, ultraviolet), that can be coupled with electrical conductivity and magnetic susceptibility sensors, to control the mechanical separation of ore into two or more categories on an individual rock by rock basis.

Dewatering

Dewatering is an important process in mineral processing. The purpose of dewatering is to remove water contained in particles. This is done for a number of reasons, specifically, to enable ore handling and concentrates to be transported easily, allow further processing to occur and to dispose of the gangue. The water removed from dewatering can be recycled through a mineral processing plant. The main processes that are used in dewatering include dewatering screens, sedimentation, filtering, and thermal drying. These processes increase in difficulty and cost as the particle size decreases. Dewatering screens operate by passing particles over a screen. The particles pass over the screen while the water passes through the apertures in the screen. This process is only viable for coarse ores that have a close size distribution as the apertures can allow small particles to pass though and are not able to be produced for small particles.

Sedimentation operates by passing water into a large thickener or clarifier. In these devices, the particles settle out of the slurry under the effects of gravity or centripetal forces. These are limited by the surface chemistry of the particles and the size of the particles. To aid in the sedimentation process, floculants and coagulants are added to reduce the repulsive forces between the particles. This repulsive force is due to the double layer formed on the surface of the particles. The floculants work by binding multiple particles together while the coagulants work by reducing the thickness of the charged layer on the outside of the particle.

Thermal drying is usually used for fine particles and to remove low water content in the particles. Some common processes include rotary dryers, fluidised beds, spray driers, hearth dryers and rotary tray dryers. This process is usually expensive to operate due to the heating requirements of the dryers.

OTHER PROCESSES

Many mechanical plants also incorporate hydrometallurgical or pyrometallurgical processes as part of an extractive metallurgical operation. Geometallurgy is a branch ofextractive metallurgy that combines mineral processing with the geologic sciences. A number of auxiliary materials handling operations are also considered a branch of mineral processing such as storage (as in bin design), conveying, sampling, weighing, slurry transport, and pneumatic transport.

FROTH FLOTATION

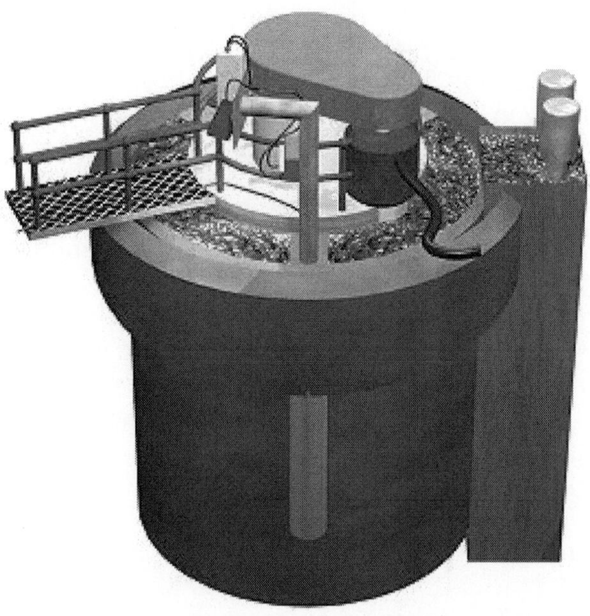

Fig. Diagram of a cylindrical froth flotation cell with camera and light used in image analysis of the froth surface.

Froth flotation is a process for selectively separating hydrophobic materials from hydrophilic. This is used in several processing industries. Historically this was first used in the mining industry, where it was one of the great enabling technologies of the 20th century. It has been described as "the single most important operation used for the recovery and upgrading of sulfide ores".

The development of froth flotation improved the recovery of valuableminerals, such as copper- and lead-bearing minerals. Along with mechanized mining, it allowed the economic recovery of valuable metals at much lower grades than before.

HISTORY

Initially, naturally occurring chemicals such as fatty acids and oils were used as flotation reagents in a large quantity to increase the hydrophobicity of the valuable minerals. Since then, the process has been adapted and applied to a wide variety of materials to be separated, and additional collector agents, including surfactants and synthetic compounds have been adopted for various applications. William Haynes patented a process in 1869 for separating sulfide and gangue minerals using oil and called it bulk-oil flotation. In 1885 Carrie Everson expanded upon this and patented a process calling for oil[s] but also an acid or a salt.

The first successful commercial flotation process for mineral sulphides was invented by Frank Elmore who worked on the development with his brother, Stanley. The Glasdir copper mine at Llanelltyd, near Dolgellau in North Wales was bought in 1896 by the Elmore brothers in conjunction with their father, William. In 1897, the Elmore brothers installed the world's first industrial size commercial flotation process for mineral beneficiation at the Glasdir mine. The process was not froth flotation but used oil to agglomerate (make balls of) pulverised sulphides and buoy them to the surface, and was patented in 1898 with a description of the process published in 1903 in theEngineering and Mining Journal.

By this time they had recognized the importance of air bubbles in assisting the oil to carry away the mineral particles. The Elmores had formed a company known as the Ore Concentration Syndicate Ltd to promote the commercial use of the process worldwide. However developments elsewhere, particularly inAustralia by Minerals Separation Ltd., led to decades of hard fought legal battles and litigations which, ultimately, were lost as the process was superseded by more advanced techniques. Charles Butters, beginning about 1899, and working with both the Elmores and Minerals Separation's representative E.H. Nutter developed what was known to contemporaries as the "Butters Process". The flotation process was independently invented in the early 1900s in Australia by Charles Vincent Potter and around the same time by Guillaume Daniel Delprat. This process (developed circa 1902) did not use oil, but relied upon flotation by the generation of gas formed by the introduction of acid into the pulp.

In 1902, Froment combined oil and gaseous flotation using a modification of the Potter-Delprat process. Another process was developed in 1902 by Cattermole, who emulsified the pulp with a small quantity of oil, subjected it to violent agitation, then slow stirring which coagulated the target minerals into nodules which were separated from the pulp by gravity. This was the basis of the Minerals Separation Ltd. process. By 1904, the MacQuisten process (a surface tension based method) was developed but this would not work when slimes were present. In 1912, James M. Hyde modified the Minerals Separation Process and installed it in the Butte and Superior Mill in Basin,

Montana, the first such installation in the USA. John M. Callow, of General Engineering of Salt Lake City, had followed flotation from technical papers and the introduction in both the Butte and Superior Mill, and at Inspiration Copper in Arizona and determined that mechanical agitation was a drawback to the existing technology. Introducing a porous brick with compressed air, and a mechanical stirring mechanism, Callow applied for a patent in 1914. This method, known as Pneumatic Flotation, was recognized to revolutionize the process of flotation concentration. The American Institute of Mining Engineers presented Callow the James Douglas Gold Medal in 1926 for his contributions to the field of flotation. In the 1960s the froth flotation technique was adapted for deinking recycled paper.

INDUSTRIES

Mining

Fig. Froth flotation to separate plastics,Argonne National Laboratory

Froth flotation is a process for separating minerals from gangue by taking advantage of differences in their hydrophobicity. Hydrophobicity differences between valuable minerals and waste gangue are increased through the use of surfactants and wetting agents. The selective separation of the minerals makes processing complex (that is, mixed) ores economically feasible. The flotation process is used for the separation of a large range of sulfides, carbonates and oxides prior to further refinement.Phosphates and coal are also upgraded (purified) by flotation technology.

Prior to 1907, practically all the copper mined in the United States came from underground vein deposits, averaging 2.5 percent copper. By 1991, the average grade of copper ore mined in the United States had fallen to only 0.6 percent. Without the flotation process, today's society, with its copper wires for electrical conduction and electrical motors, would not have happened. Copper would be too expensive.[citation needed]

Waste water treatment

The flotation process is also widely used in industrial waste water treatment plants, where it removes fats, oil, grease and suspended solids from waste water. These units are called dissolved air flotation (DAF) units. In particular, dissolved air flotation units are used in removing oil from the wastewater effluents of oil refineries, petrochemical and chemical plants, natural gas processing plants and similar industrial facilities.

Paper recycling

Froth flotation is one of the processes used to recover recycled paper. In the paper industry this step is called deinking or just flotation. The target is to release and remove the hydrophobic contaminants from the recycled paper. The contaminants are mostlyprinting ink and stickies. Normally the setup is a two stage system with 3,4 or 5 flotation cells in series.

PRINCIPLE OF OPERATION

Before froth flotation can work, the ore to be treated is reduced to fine particles by crushing and grinding (a process known as comminution) so that the various minerals exist as physically separate grains. This process is known as liberation. The particle sizes are typically less than 0.1 mm (100 μm), but sometimes sizes smaller than 7–10 μm are required. There is a tendency for the liberation size of the minerals to decrease over time as the ore bodies with coarse mineral grains that can be separated at larger sizes are depleted and replaced by ore bodies that were formerly considered too difficult.

In the mining industry, the plants where flotation is undertaken to concentrate ore are generally known as concentrators or mills. For froth flotation, the ground ore is mixed with water to form a slurry and the desired mineral is rendered hydrophobic by the addition of a surfactant or collector chemical (although some mineral surfaces are naturally hydrophobic, requiring little or no addition of collector).

The particular chemical depends on the nature of the mineral to be recovered and, perhaps, the natures of those that are not wanted. As an example, sodium ethyl xanthate ("SEX") may be added as a collector in the selective flotation ofgalena (lead sulfide) to separate it from sphalerite (zinc sulfide). This slurry (more properly called the pulp) of hydrophobic particles and hydrophilic particles is then introduced to tanks known as flotation cells that are aerated to produce bubbles. The hydrophobic particles attach to the air bubbles, which rise to the surface, forming a froth. The froth is removed from the cell, producing a concentrate ("con") of the target mineral. Frothing agents, known as frothers, may be introduced to the pulp to promote the formation of a stable froth on top of the flotation cell.

The minerals that do not float into the froth are referred to as the flotation tailings or flotation tails. These tailings may also be subjected to further stages of flotation to recover the valuable particles that did not float the first time. This is known as scavenging. The final tailings after scavenging are normally pumped for disposal as mine fill or to tailings disposal facilities for long-term storage.

Froth flotation efficiency is determined by a series of probabilities: those of particle–bubble contact, particle–bubble attachment, transport between the pulp and the froth, and froth collection into the product launder. In a conventional mechanically-agitated cell, the void fraction (*i.e.* volume occupied by air bubbles) is low (5 to 10 percent) and the bubble size is usually greater than 1 mm. This results in a relatively low interfacial area and a low probability of particle–bubble contact. Consequently, several cells in series are required to increase the particle residence time, thus increasing the probability of particle–bubble contact. Flotation is normally undertaken in several stages to maximize the recovery of the target mineral or minerals and the concentration of those minerals in the concentrate, while minimizing the energy input.

Flotation stages

Roughing

The first stage is called roughing, which produces a rougher concentrate. The objective is to remove the maximum amount of the valuable mineral at as coarse a particle size as practical. The finer an ore is ground, the greater the energy that is required, so it makes sense to only grind fine those particles that need fine grinding. Complete liberation is not required for rougher flotation, only sufficient liberation to release enough gangue from the valuable mineral to get a high recovery.

The primary objective of roughing is to recover as much of the valuable minerals as possible, with less emphasis on the quality of the concentrate produced. In some concentrators, there may be a preflotation step that precedes roughing. This is done when there are some undesirable materials, such as organic carbon, that readily float. They are removed first to avoid them floating during roughing (and thus contaminating the rougher concentrate).

Cleaning

The rougher concentrate is normally subjected to further stages of flotation to reject more of the undesirable minerals that also reported to the froth, in a process known ascleaning. The product of cleaning is known as the cleaner concentrate or the final concentrate.

The objective of cleaning is to produce as high a concentrate grade as possible. The rougher concentrate is often subject to further grinding (usually called regrinding) to get more complete liberation of the valuable minerals. Because it is a smaller mass than that of the original ore, less energy is needed than would be necessary if the whole ore were reground. Regrinding is often undertaken in specialized regrind mills, such at the IsaMillTM, designed to further reduce the energy consumed during regrinding to finer sizes.

Scavenging

The rougher flotation step is often followed by a scavenger flotation step that is applied to the rougher tailings. The objective is to recover any of the target minerals that were not recovered during the initial roughing stage. This might be achieved by changing the flotation conditions to make them more rigorous than the initial roughing, or there might be some secondary grinding to provide further liberation.

The concentrate from the rougher scavengers could be returned to the rougher feed for refloating or Similarly, the cleaning step may also be followed by a scavenging step performed on the cleaner tailings.

SCIENCE OF FLOTATION

To be effective on a given ore slurry, the collectors are chosen based upon their selective wetting of the types of particles to be separated. A good collector will adsorb, physically or chemically, with one of the types of particles. This provides the thermodynamic requirement for the particles to bind to the surface of a bubble. The wetting activity of a surfactant on a particle can be quantified by measuring the contact angles that the liquid/bubble interface makes with it. Another important measure for attachment of bubbles to particles is induction time. The induction time is the time required for the particle and bubble to rupture the thin film separating the particle and bubble. This rupturing is achieved by the surface forces between the particle and bubble.

The mechanisms for the bubble-particle attachment is very complex and consists of three steps, collision, attachment and detachment. The collision is achieved by particles being within the collision tube of a bubble and this is affected by the velocity of the bubble and radius of the bubble. The collision tube corresponds to the region in which a particle will collide with the bubble, with the perimeter of the collision tube corresponding to the grazing trajectory. The attachment of the particle to the bubble is controlled by the induction time of the particle and bubble. The particle and bubble need to bind and this occurs if the time in which the particle and bubble are in contact with each other is larger than the required induction time. This induction time is effected by the fluid viscosity, particle and bubble size and the forces between the particle and bubbles.

The detachment of a particle and bubble occurs when the force exerted by the surface tension is exceeded by shear forces and gravitational forces. These forces are complex and vary within the cell. High shear will be experienced close to the impeller of a mechanical flotation cell and mostly gravitational force in the collection and cleaning zone of a flotation column. Significant issues of entrainment of fine particles occurs as these particles experience low collision efficiencies as well as sliming and degradation of the particle surfaces. Coarse particles show a low recovery of the valuable mineral due to the low liberation and high detachment efficiencies.

FLOTATION EQUIPMENT

Flotation can be performed in rectangular or cylindrical mechanically agitated cells or tanks, flotation columns, Jameson Cell or deinking flotation machines. Comparison of the sizes of conventional flotation columns andJameson Cells with similar capacities. Mechanical cells use a large mixer and diffuser mechanism at the bottom of the mixing tank to introduce air and provide mixing action.

Flotation columns use airspargers to introduce air at the bottom of a tall column while introducing slurry above. The countercurrent motion of the slurry flowing down and the air flowing up provides mixing action. Mechanical cells generally have a higher throughput rate, but produce material that is of lower quality, while flotation columns generally have a low throughput rate but produce higher quality material. The Jameson cell uses neither impellers nor spargers, instead combining the slurry with air in a downcomer where high shear creates the turbulent conditions required for bubble particle contacting.

MECHANICS OF FLOTATION

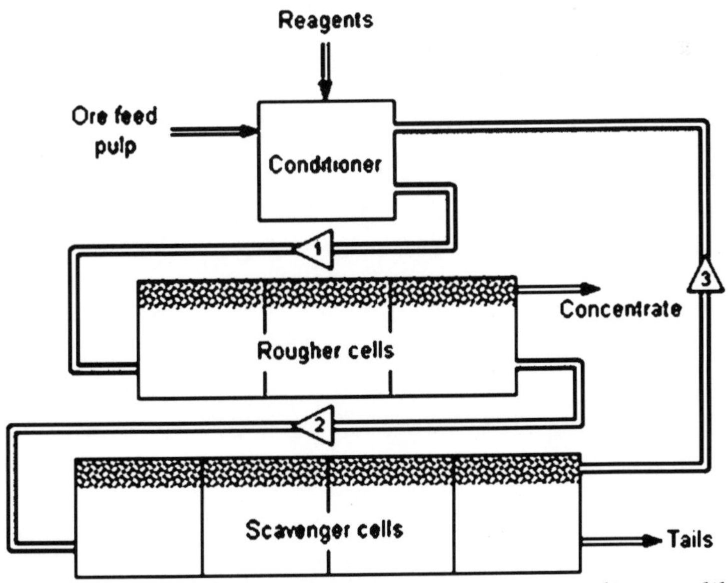

The following steps are followed, following grinding to liberate the mineral particles.

1. Reagent conditioning to achieve hydrophobic surface charges on the desired particles
2. Collection and upward transport by bubbles in an intimate contact with air or nitrogen
3. Formation of a stable froth on the surface of the flotation cell
4. Separation of the mineral laden froth from the bath (flotation cell)

Simple flotation circuit for mineral concentration. Numbered triangles show direction of stream flow, Various flotation reagents are added to a mixture of ore and water (called pulp) in a conditioning tank. The flow rate and tank size are designed to give the minerals enough time to be activated. The conditioner pulp is fed to a bank of rougher cells which remove most of the desired minerals as a concentrate.

The rougher pulp passes to a bank of scavenger cells where additional reagents may be added. The scavenger cell froth is usually returned to the rougher cells for additional treatment, but in some cases may be sent to special cleaner cells. The scavenger pulp is usually barren enough to be discarded as tails. More complex flotation circuits have several sets of cleaner and re-cleaner cells, and intermediate re-grinding of pulp or concentrate.

CHEMICALS OF FLOTATION

Collectors[edit]

Collectors either chemically bond (chemisorption) on a hydrophobic mineral surface, or adsorb onto the surface in the case of, for example, coal flotation through physisorption. Collectors increase the natural hydrophobicity of the surface, increasing theseparability of the hydrophobic and hydrophilic particles.

- Xanthates
- Potassium amyl xanthate (PAX)
- Potassium isobutyl xanthate (PIBX)
- Potassium ethyl xanthate (KEX)
- Sodium isobutyl xanthate (SIBX)
- Sodium isopropyl xanthate (SIPX)
- Sodium ethyl xanthate (SEX)

Dithiophosphates:

- Thiocarbamates
- Xanthogen Formates
- Thionocarbamates
- Thiocarbanilide.

Frothers:

- Pine oil
- Alcohols (methyl isobutyl carbinol (MIBC))
- Polyglycols

- Polyoxyparafins l
- Cresylic Acid (Xylenol)

Modifiers

pH modifiers such as:

- Lime CaO
- Soda ash Na_2CO_3
- Caustic soda NaOH
- Acid H_2SO_4, HCl

Cationic modifiers:

- Ba^{2+}, Ca^{2+}, Cu+, Pb^{2+}, Zn^{2+}, Ag+

Anionic modifiers:

- SiO_3^{2-}, PO_4^{3-}, CN-, CO_3^{2-}, S^{2-}

Organic modifiers:

- Dextrin, starch, glue, CMC

Chemicals for deinking of recycled paper[edit]

- pH control: sodium silicate and sodium hydroxide
- Calcium ion source: hard water, lime or calcium chloride
- Collector: fatty acid, fatty acid emulsion, fatty acid soap and/or organo-modified siloxane

TROMMEL

A trommel (from the German word for drum, Trommel)[citation needed] is a screened cylinder used to separate materials by size - for example, separating the biodegradable fraction of mixed municipal waste or separating different sizes of crushedstone. Portable trommels (also called portable trommel screens) are often used in the production of organic products from various types of waste. For example, excavation contractors may screen their site debris into two fractions; a saleable topsoil for farms, nurseries and site-work, as well as cleaned rock for aggregates or landscaping work. This allows the contractor to resell their waste, instead of incurring the cost of sending it for disposal.

Fig. An operating trommel processing mixed waste

Fig. Portable trommel screening topsoil

The same principle applies to the production of compost, sand/gravel, lumber mill by-products and municipal waste.

Trommels are also used in the recovery of Gold from mineral rich soils. The dirt is broken up by high pressure water jets before entering the trommel which breaks up the dirt and outputs the rocks and stones while depositing the gold fragments in special plastic mats.

GOLD PANNING

Gold panning, or simply panning, is a form of placer mining and traditional mining that extracts gold from a placer deposit using a pan. The process is one of the simplest ways to extract gold, and is popular with geology enthusiasts especially because of its cheap cost and the relatively simple and easy process.

Fig. Panning for gold in a creek bed

It is the oldest method of mining gold. The first recorded instances of placer mining are from ancient Rome, where gold and other precious metals were extracted from streams and mountainsides using sluices and panning.

However, the productivity rate is comparably smaller compared to other methods such as the rocker box or a large extractor such as in 'The Pit', in Kargoorlie, which have led to it largely being replaced in areas with capital for more sophisticated equipment.

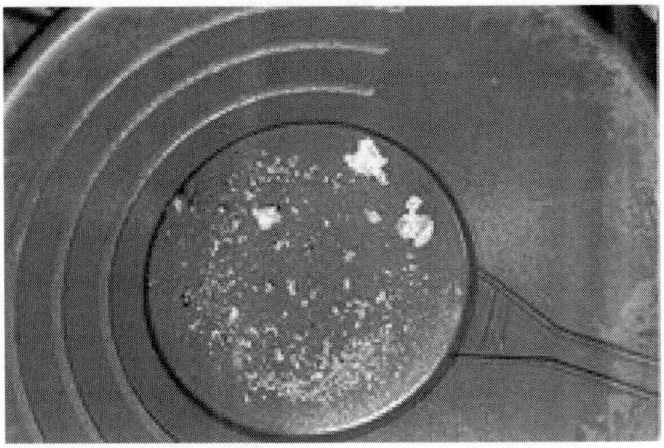

Fig. Gold in the pan, Alaska

PROCESS

Gold panning is a simple process. Once a suitable placer deposit is located, some gravel from it is scooped into a pan, where it is then gently agitated in water and the gold sinks to the bottom of the pan. Materials with a low specific gravity are allowed to spill out of the pan, whereas materials with a high specific gravity sink to the bottom of the sediment during agitation and remain within the pan for examination and collection by the gold panner. These dense materials usually consist primarily of a black, magnetite sandwith whatever gemstones or metal dust that may be found in the deposit that is used for source material.

While an effective method with certain kinds of deposits, and essential for prospecting, even skilled panners can work but a limited amount of material, significantly less than the other methods which have replaced it in larger operation. Pans remain in use in places where there is limited capital or infrastructure, as well as in recreational gold mining.

In many situations, gold panning usually turns up only minor gold dust that is usually collected as a souvenir in small clear tubes. Nuggets and considerable amounts of dust are occasionally found, but panning mining is not generally lucrative. Panning for gold can be used to locate the parent gold veins which are the source of most placer deposits.

PANS

Fig. Various designs of gold pans from around the world

Gold pans of various designs have been developed over the years, the common features being a means for trapping the heavy materials during agitation, or for easily removing them at the end of the process. Some are intended for use with mercury, include screens, sharp corners for breaking ice, are non-round, or are even designed for use "with or without water". Edward Otho Cresap Ord, II, a former Army officer and co-owner of several mines, patented several pan designs including designs for use with mercury or dry. Pans are measured by their diameter in inches or centimeters. Common sizes of gold pans today ranges between 10–17 inches (25–43 cm), with 14 inches (36 cm) being the most used size. The sides are generally angled between 30° to 45°. Pans are manufactured in both metal and high impact plastic. Russia iron or heavy gauge steel pans are traditional. Steel pans are heavier and stronger than plastic pans. Some are made of lightweight alloys for structural stability. Plastic gold pans resist rust, acid and corrosion, and most are designed with moulded riffles along one side of the pan. Of the plastic gold pans, green and red ones are usually preferred among prospectors, as both the gold and the black sand stands out in the bottom of the pan.

Batea

The batea (Spanish for "gold pan") is a particular variant of gold pan. Traditionally made of a solid piece of wood, it may also be made of metal. Bateas are used in areas where there is less water available for use than with traditional gold pans, such as Mexico and South America, where it was introduced by the Spanish. Bateas are larger than other gold pans, being closer to 20 inches (51 cm) in diameter.

ROCKER BOX

A rocker box (also known as a cradle) is a gold mining implement for separating alluvial placer gold from sand and gravel which was used in placer mining in the 19th century. It consists of a high-sided box, which is open on one end and on top, and was placed on rockers.

The inside bottom of the box is lined with riffles and a usually a carpet similar to a sluice. On top of the box is a classifier sieve (usually with half-inch or quarter-inch openings) which screens-out larger pieces of rock and other material, allowing only finer sand and gravel through. Between the sieve and the lower sluice section is a baffle, which acts as another trap for fine gold and also ensures that the aggregate material being processed is evenly distributed before it enters the sluice section. It sits at an angle and points towards the closed back of the box.

Traditionally, the baffle consisted of a flexible apron or "blanket riffle" made of canvas or a similar material, which had a sag of about an inch and a half in the center, to act as a collection pocket for fine gold. Later rockers (including most modern ones) dispensed with the flexible apron and used a pair of solid wood or metal baffle boards which. These are sometimes covered with carpet to trap fine gold. The entire device sits on rockers at a slight gradient, which allows it to be rocked side to side.

Today, the rocker box is not used as extensively as the sluice, but still is an effective method of recovering gold in areas where there is not enough available water to operate a sluice effectively. Like a sluice box, the rocker box has riffles and a carpet in it to trap the gold. It was designed to be used in areas with less water than a sluice box. The process involves pouring water out of a small cup and then rocking the small sluice box like a cradle, thus the name rocker box or cradle.

Rocker boxes must be manipulated carefully, to prevent losing the gold. Although big, and difficult to move, the rocker can pick up twice the amount of the gravel, and therefore more gold in one day than an ordinary gold mining pan. The rocker, like the pan, is used extensively in small-scale placer work, in sampling, and for washing sluice concentrates and material cleaned by hand from bedrock in other placer operations. One to three cubic yards, bank measure, can be dug and washed in a rocker per man-shift, depending upon the distance the gravel or water has to be carried, the character of the gravel, and the size of the rocker.

Rockers are usually homemade and display a variety of designs. A favorite design consists essentially of a combination washing box and screen, a canvas or carpet apron under the screen, a short sluice with two or more riffles, and rockers under the sluice. The bottom of the washing box consists of sheet metal with holes about 1/2 inch in diameter punched in it, or a 1/2-inch-mesh screen can be used. Dimensions shown are satisfactory but variations are possible. The bottom of the rocker should be made of a single wide, smooth board, which will greatly facilitate cleanups. The materials for building a rocker cost only a few dollars, depending mainly upon the source of lumber.

DRYWASHER

The drywasher is a common desert mining tool for gold mining. The drywasher is like a highbanker since it uses a motor and a form of sluice however it lacks the need for water. A drywasher operates by the use of air. By forcing air to flow up through the material as it moves down the sluice the heavier materials, like gold, will stay at the bottom and get trapped by the riffles while the lighter materials will flow up and over the riffles or be blown away. The lighterweight material that is not blown away will slowly flow down the sluice and fall out the end of the sluice.

Drywasher - A mechanical apparatus used for separating free particles of placer gold from dry sediments. Dry placers represent a major portion of all alluvial gold deposits around the globe. Extreme aridity in southern hemispheric regions of the planet causes a slower rate of erosion on placer deposits than if rainfall were plentiful and regional watersheds were hydrologically active. This absence of water directly correlates to the low moisture content of alluvial sediments and the short distance gold moves from its hardrock source. A drywasher works on the principle of gravity separation. Since gold, at 19.1 specific gravity, is much heavier than most other minerals found in local sediments, a facsimile of certain natural conditions (which would involve the use of water, if it were available) can be mechanically reproduced with a simple machine using dry air instead of water as the medium of separation.

The process of excavating dry, compacted sediments will often loosen or unconsolidate them. Then, as long as the material is sufficiently dry, these gravels can be processed with a drywasher in a relatively efficient manner. A drywasher was also known, at one time, as a Mexican air jig. To use a Drywasher, first, you determine if the material you want to run is dry enough to be efficiently processed. Then, set the machine up and stabilize it. Once it's securely positioned, you transfer (shovel or load) the dirt, clay, sand, gravel, et al. onto the uppermost part of the device.....the 'grizzly'. Sediments classified by this screen (or grizzly), drop into another part of the apparatus known as 'the hopper'.

Once the machine is activated, the reduced size, gold-bearing material is then agitated downward, at an angle, pulsing over a series of "riffles" in a recovery tray. This process varies ever so slightly from machine to machine. Heavier minerals, principally gold and fine-grained black sands (various iron minerals) are separated from the paydirt and reconcentrated by this means. Panning, or some other final separation process, comes afterwards. Drywashers are a fairly modern invention. It is doubtful anyone knows for sure when early prototypes were first employed in the goldfields, but these primitive machines were, in all likelihood, operated manually (by hand). A simple one can be constructed using a wooden framework, a hopper, a recovery tray, bellows, lever, and cord. Before the advent of modern metal screening material, sorting of gravels into smaller particles might have been done by hand. Such elimination of larger rocks aids in separating gold from dry, auriferous sediments to be processed, consequently, any modern 'puffer-belly' could be outfitted with a screen, or 'grizzly' for sorting the dry gravels into two sizes.

EXTRACTIVE METALLURGY

Extractive metallurgy is the study of the processes used in the separation and concentration (beneficiation) of raw materials. The field is an applied science, covering all aspects of the physical and chemical processes used to produce mineral-containing and metallic materials, sometimes for direct use as a finished product, but more often in a form that requires further physical processing which is generally the subject of physical metallurgy, ceramics, and other disciplines within the broad field of materials science.

The field of ferrous and non-ferrous extractive metallurgy of encompasses many specialty sub-disciplines, each concerned with various physical and chemical processes that are steps in an overall process of producing a particular material. These specialties are generically grouped into the categories of mineral processing, hydrometallurgy,pyrometallurgy, and electrometallurgy. The distinctions among these groups of sub-disciplines within extractive metallurgy is far from clear, and many commercially important metallurgical processes involve considerable overlap.

The theoretical basis of extractive metallurgy is underpinned by the more general sciences of physics, chemistry, and geology. Additionally, the practice of extractive metallurgy nearly always involves contributions from other scientific fields such as analytical chemistry and mineralogy. Extractive metallurgy is the practice of removing valuable metals from an ore and refining the extracted raw metals into a purer form. In order to convert a metal oxide or sulfide to a purer metal, the ore must be reducedphysically, chemically, or electrolytically.

Extractive metallurgists are interested in three primary streams: feed, concentrate (valuable metal oxide/sulfide), and tailings (waste). After mining, large pieces of the ore feed are broken through crushing and/or grinding in order to obtain particles small enough where each particle is either mostly valuable or mostly waste. Concentrating the particles of value in a form supporting separation enables the desired metal to be removed from waste products.

Mining may not be necessary if the ore body and physical environment are conducive to leaching. Leaching dissolves minerals in an ore body and results in an enriched solution. The solution is collected and processed to extract valuable metals. Ore bodies often contain more than one valuable metal. Tailings of a previous process may be used as a feed in another process to extract a secondary product from the original ore. Additionally, a concentrate may contain more than one valuable metal. That concentrate would then be processed to separate the valuable metals into individual constituents.

MINERAL PROCESSING

Mineral processing involves the processes used to manipulate the particle size of solid raw materials and to separate valuable materials from materials of no value, referred to as gangue. Usually, particle size reduction, also referred to as comminution, is required to permit efficient separation of the valuable materials from gangue.

Separation processes take advantage of physical properties of the materials in order to separate them from each other. These physical properties can include density, particle size and shape, electrical and magnetic properties, and surface properties. Since many size reduction and separation processes involve the use of water, solid-liquid separation processes are also a subject of mineral processing.

Hydrometallurgy

Hydrometallurgy is concerned with processes involving aqueous solutions to extract metals from ores. The most common hydrometallurgical process is leaching, which involves dissolution of the valuable metals into the aqueous solution. After the solution is separated from the ore solids, the solution is often subjected to various processes of purification and concentration before the valuable metal is recovered either in its metallic state or as a chemical compound.

The solution purification and concentration processes may include precipitation, distillation, adsorption, and solvent extraction. The final recovery step may involve precipitation, cementation, or an electrometallurgical process. Sometimes, hydrometallurgical processes may be carried out directly on the ore material without any pretreatment steps. More often, the ore must be pretreated by various mineral processing steps, and sometimes by pyrometallurgical processes.

Pyrometallurgy

Pyrometallurgy involves high temperature processes where chemical reactions take place among gases, solids, and molten materials. Solids containing valuable metals are reacted to form intermediate compounds for further processing or converted into their elemental or metallic state. Pyrometallurgical processes that involve gases and solids are typified by calcining and roastingoperations. Processes that produce molten products are collectively referred to as smelting operations. The energy required to sustain the high temperature pyrometallurgical processes may come entirely from the exothermic nature of the chemical reactions taking place, usually oxidation reactions. Often, however, energy must be added to the process by combustion of fuel or, in the case of some smelting processes, by the direct application of electrical energy. Ellingham Diagrams are a useful way of analysing the possible reactions, and so predicting their outcome.

ELECTROMETALLURGY

Electrometallurgy involves metallurgical processes that take place in some form of electrolytic cell. The most common types of electrometallurgical processes are electrowinning and electro-refining. Electrowinning is an electrolysis process used to recover metals in aqueous solution, usually as the result of an ore having undergone one or more hydrometallurgical processes. The metal of interest is plated onto the cathode, while the anode is an inert electrical conductor.

Electro-refining is used to dissolve an impure metallic anode (typically from a smelting process) and produce a high purity cathode. Fused salt electrolysis is another electrometallurgical process whereby the valuable metal has been dissolved into a molten salt which acts as the electrolyte, and the valuable metal collects on the cathode of the cell. The fused salt electrolysis process is conducted at temperatures sufficient to keep both the electrolyte and the metal being produced in the molten state. The scope of electrometallurgy has significant overlap with the areas of hydrometallurgy and (in the case of fused salt electrolysis) pyrometallurgy. Additionally, electrochemical phenomena play a considerable role in many mineral processing and hydrometallurgical processes.

PYROMETALLURGY

Pyrometallurgy is a branch of extractive metallurgy. It consists of the thermal treatment of minerals and metallurgical ores and concentrates to bring about physical and chemical transformations in the materials to enable recovery of valuable metals. Pyrometallurgical treatment may produce saleable products such as pure metals, or intermediate compounds or alloys, suitable as feed for further processing. Examples of elements extracted by pyrometallurgical processes include the oxides of less reactive elements like Fe, Cu, Zn, Chromium, Tin, Manganese.

Pyrometallurgical processes are generally grouped into one or more of the following categories:

- Calcining
- Roasting
- Smelting
- Refining

Most pyrometallurgical processes require energy input to sustain the temperature at which the process takes place. The energy is usually provided in the form of fossil fuel combustion, exothermic reaction of the material, or from electrical heat. When enough material is present in the feed to sustain the process temperature solely by exothermic reaction (*i.e.* without the addition of fuel or electrical heat), the process is said to be "autogenous."

CALCINING

Calcining is thermal decomposition of a material. Examples include decomposition of hydrates such as ferric hydroxide to ferric oxide and water vapor, or decomposition of calcium carbonate to calcium oxide and carbon dioxide and or of iron carbonate to iron oxide. Calcination processes are carried out in a variety of furnaces, including shaft furnaces, rotary kilns, and fluidized bed reactors.

ROASTING

Roasting consists of thermal gas-solid reactions, which can include oxidation, reduction, chlorination, sulfation, and pyrohydrolysis. The most common example of roasting is the oxidation of metal sulfide ores. The metal sulfide is heated in the presence of air to a temperature that allows the oxygen in the air to react with the sulfide to form sulfur dioxide gas and solid metal oxide. The solid product from roasting is often called "calcine." In oxidizing roasting, if the temperature and gas conditions are such that the sulfide feed is completely oxidized, the process is known as "dead roasting."

Sometimes, as in the case of pre-treating reverberatory or electric smelting furnace feed, the roasting process is performed with less than the required amount of oxygen to fully oxidize the feed. In this case, the process is called "partial roasting," because the sulfur is only partially removed. Finally, if the temperature and gas conditions are controlled such that the sulfides in the feed react to form metal sulfates instead of metal oxides, the process is known as "sulfation roasting." Sometimes, temperature and gas conditions can be maintained such that a mixed sulfide feed (for instance a feed containing both copper sulfide and iron sulfide) reacts such that one metal forms a sulfate and the other forms an oxide, the process is known as "selective roasting" or "selective sulfation."

SMELTING

Smelting involves thermal reactions in which at least one product is a molten phase. Metal oxides can then be smelted by heating with coke or charcoal (forms of carbon), a reducing agent that liberates the oxygen as carbon dioxide leaving a refined mineral. Concern about the production of carbon dioxide is only a recent worry, following the identification of the enhanced greenhouse effect.

Carbonate ores are also smelted with charcoal, but sometimes need to be calcined first.[citation needed] Other materials may need to be added as flux, aiding the melting of the oxide ores and assisting in the formation of a slag, as the flux reacts with impurities, such as siliconcompounds.[citation needed] Smelting usually takes place at a temperature above the melting point of the metal, but processes vary considerably according to the ore involved and other matters.[citation needed]

REFINING

Refining is the removal of impurities from materials by a thermal process. This covers a wide range of processes, involving different kinds of furnace or other plant. The term, 'refining' can also refer to certain electrolytic processes. Accordingly, some kinds of pyrometallurgical refining are referred to as 'fire refining'.

CUPELLATION

Cupellation is a refining process in metallurgy, where ores or alloyed metals are treated under high temperatures and controlled operations to separate noble metals, like gold and silver, from base metals like lead, copper, zinc, arsenic, antimony or bismuth, present in the ore.

The process is based on the principle that precious metals do not oxidise or react chemically, unlike the base metals; so when they are heated at high temperatures, the precious metals remain apart and the others react forming slags or other compounds.

Since the Early Bronze Age, the process was used to obtain silver from smelted lead ores. By the Middle Ages and the Renaissance, cupellation was one of the most common processes for refining precious metals. By then, fire assays were used for assaying minerals, that is, testing fresh metals such as lead and recycled metals to know their purity for jewellery and coin making. Cupellation is still in use today.

PROCESS

Large scale cupellation

Native silver is a rare element, although it exists as such. It is usually found in nature combined with other metals, or in minerals that contain silver compounds, generally in the form of sulfides such as galena (lead sulfide) or cerussite (lead carbonate). So the primary production of silver requires the smelting and then cupellation of argentiferous lead ores. Lead melts at 327°C while silver melts at 960°C; when mixed, as in galena, the most common argentiferous lead ore, they have to be smelted at high temperatures in areducing atmosphere to produce argentiferous lead.

The alloy is melted again at the high temperature of 900°C to 1000°C in a hearth or blast furnace, where air flow makes possible the oxidation of the lead. The lead oxidises to lead oxide (PbO) known as litharge, which captures the oxygen from the other metals present, while silver and gold remain on top of the liquid litharge. The latter is removed or absorbed by capillary action into the hearth linings.

This chemical reaction may be viewed as :

$$(Ag+Cu) + Pb + O2 \rightarrow (CuO+PbO) + Ag$$

The base of the hearth was dug in the form of a saucepan, and covered with an inert and porous material rich in calcium or magnesium such as shells, lime, or the ash from burning wood or bones. The lining had to be calcareous because lead reacts with silica (clay compounds) to form viscous lead silicate that prevents the needed absorption of litharge, whereas calcareous materials do not react with lead. Some of the litharge evaporates, and the rest is absorbed by the porous earth lining to form "litharge cakes".

Litharge cakes are usually circular or concavo-convex, about 15 cm in diameter. They are the most common archaeological evidence of cupellation in the Early Bronze Age. By their chemical composition, archaeologists can tell what kind of ore was treated, its main components, and the chemical conditions used in the process. This permits insights about production process, trade, social needs or economic situations.

Small scale cupellation

Small scale cupellation is based on the same principle as the one done in a cupellation hearth; the main difference lies in the amount of material to be tested or obtained. The minerals have to be crushed, roasted and smelted to concentrate the metallic components in order to separate the noble metals. By the Renaissance the use of the cupellation processes was diverse: assay of ores from the mines, testing the amount of silver in jewels or coins or for experimental purposes. It was carried out in small shallow recipients known as cupels.

As the main purpose of small scale cupellation was to assay and test minerals and metals, the matter to be tested has to be carefully weighed. The assays were made in the cupellation or assay furnace, which needs to have windows and bellows to ascertain that the air oxidises the lead, as well as to be sure and prepared to take away the cupel when the process is over. Pure lead has to be added to the matter being tested to guarantee the further separation of the impurities. After the litharge has been absorbed by the cupel, buttons of silver were formed and settled in the middle of the cupel. If the alloy also contained a certain amount of gold, it settled with the silver and both had to be separated by parting.

Cupels

The primary tool for small scale cupellation was the cupel. Cupels were manufactured in a very careful way. They used to be small vessels shaped in the form of an inverted truncated cone, made out of bone ashes. According to Georg Agricola, the best material was obtained from burned horns of deer although fish spines could work well. Ashes have to be ground into a fine and homogeneous powder and mixed with some sticky substance to mould the cupels. Moulds were made out of brass with no bottoms so that the cupels could be taken off. A shallow depression in the centre of the cupel was made with a rounded pestle. Cupel sizes depend on the amount of material to be assayed. This same shape has been maintained until the present.

Archaeological investigations as well as archaeometallurgical analysis and written texts from the Renaissance have demonstrated the existence of different materials for their manufacture; they could be made also with mixtures of bones and wood ashes, of poor quality, or moulded with a mixture of this kind in the bottom with an upper layer of bone ashes. Different recipes depend on the expertise of the assayer or on the special purpose for which it was made (assays for minting, jewelry, testing purity of recycled material or coins). Archaeological evidence shows that at the beginnings of small scale cupellation, potsherds or clay cupels were used.

HISTORY

The first known use of silver was in the Near East in Anatolia and Mesopotamia during the 4th and 3rd millennium BC., the Early Bronze Age. Archaeological findings of silver and lead objects together with litharge pieces and slag have been studied in a variety of sites, and metallurgical analysis suggests that by then people were confidently extracting silver from lead ores so the method would have been known earlier.

During the following Iron Age, cupellation was done by fusing the debased metals with a surplus of lead, the bullion or result product of this fusion was then heated in a cupellation furnace to separate the noble metals. Mines such as Rio Tinto, near Huelva in Spain, started to be an important political and economic site for many people around the Mediterranean Sea, as well as Laurion in Greece. Around 500 B.C. control over the Laurion mines gave Athens political advantage and power in the Mediterranean so that they were able to defeat the Persians.

During the Roman times, the empire needed large quantities of lead in order to support the Roman civilization over a great territory; they searched for open lead-silver mines in whatever area they conquered. Silver coinage became the normalised media for exchange, therefore silver production and mine control gave economic and political power. In Roman times it was worth mining lead ores if their content of silver was 0.01% or more.

The origin of the use of cupellation for analysis is not known. One of the earliest written references to cupels is Theophilus Divers Ars in the 12th century AD. The process then changed little until the 16th century.

Small-scale cupellation may be considered the most important fire assay developed in history, and perhaps the origin of chemical analysis. Most of the written evidence comes from the Renaissance in the 16th century. Georg

Agricola and Lazarus Ercker, among others, wrote about the art of mining and testing the ores, as well as detailed descriptions of cupellation. Their descriptions and assumptions have been identified within diverse archaeological findings through Medieval and Renaissance Europe. By these times the amount of fire assays increased considerably, mainly because of testing the ores in the mines in order to identify the availability of its exploitation.

A primary use of cupellation was related to minting activities, and it was also used in testing jewelry. Since the Renaissance, cupellation became a standardised method of analysis that has changed very little, demonstrating its efficiency. Its development certainly touched the spheres of economy, politics, warfare and power in ancient times.

NEW WORLD

The huge amount of Pre-Hispanic silver adornments known especially from Perú, Bolivia and Ecuador make us wonder whether the pre-Hispanic civilizations obtained the raw material from native ores or from argentiferous-lead ores. Although native silver may be available in America, it is as rare as in the Old World. From colonial texts it is known that silver mines were open in colonial times by the Spaniards from Mexico to Argentina, the main ones, being those of Tasco, Mexico and Potosí in Bolivia.

Some kind of blast furnaces called huayrachinas were described in colonial texts, as native technology furnaces used in Perú and Bolivia to smelt the ores that come from the silver mines owned by the Spaniards. Although it is not conclusive, it is believed that these kinds of furnaces may have been used before the Spanish Conquest.Ethnoarchaeological and archaeological works done at Porco – Potosí, Bolivia have suggested it.

There are no specific archaeological accounts about silver smelting or mining in the Andes prior to the Incas. However, silver and lead artefacts have been found in the Peruvian central highlands dated in the pre-Inca and Inca periods. From the presence of lead in silver artefacts, archaeologists suggest that cupellation may have occurred there.

HYDROMETALLURGY

Hydrometallurgy is a method for obtaining metals from their ores. It is a technique within the field of extractive metallurgy involving the use of aqueous chemistry for the recovery of metals from ores, concentrates, and recycled or residual materials. The complement to hydrometallurgy is pyrometallurgy, which is usually older technology. Hydrometallurgy is typically divided into three general areas:

- Leaching
- Solution concentration and purification
- Metal recovery

LEACHING

Leaching involves the use of aqueous solutions containing a lixiviant which is brought into contact with a material containing a valuable metal. The lixiviant in solution may beacidic or basic in nature.

The type and concentration of the lixiviant is normally controlled to allow some degree of selectivity for the metal or metals that are to be recovered. In the leaching process, oxidation potential, temperature, and pH of the solution are important parameters, and are often manipulated to optimize dissolution of the desired metal component into the aqueous phase. The three basic leaching techniques are in-situ leaching, heap leaching, and vat leaching.

In-situ leaching

In-situ leaching is also called "solution mining." The process initially involves drilling of holes into the ore deposit. Explosives or hydraulic fracturing are used to create open pathways within the deposit for solution to penetrate into. Leaching solution is pumped into the deposit where it makes contact with the ore. The solution is then collected and processed. The Beverley uranium deposit is an example of in-situ leaching.[citation needed]

Heap leaching

In heap leaching processes, crushed (and sometimes agglomerated) ore is piled in a heap which is lined with an impervious layer. Leach solution is sprayed over the top of the heap, and allowed to percolate downward through the heap. The heap design usually incorporates collection sumps which allow the "pregnant" leach solution (*i.e.* solution with dissolved valuable metals) to be pumped for further processing.

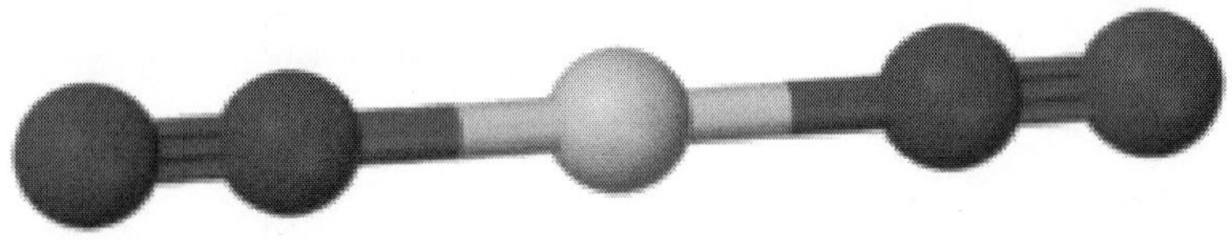

Fig. Ball-and-stick model of the aurocyanide or dicyanoaurate(I) complex anion, [Au(CN)2]-.

Vat leaching

Vat leaching involves contacting material, which has usually undergone size reduction and classification, with leach solution in large tanks or vats. Often the vats are equipped with agitators to keep the solids in suspension in the vats and improve the solid to liquid contact. After vat leaching, the leached solids and pregnant solution are usually separated prior to further processing.[citation needed]

SOLUTION CONCENTRATION AND PURIFICATION

After leaching, the leach liquor must normally undergo concentration of the metal ions that are to be recovered. Additionally, undesirable metal ions sometimes require removal.

- Precipitation
- Cementation
- Solvent Extraction
- Ion Exchange

Solvent extraction

A mixture of an extractant in a diluent is used to extract a metal from one phase to another. In solvent extraction this mixture is often referred to as the "organic" because the main constituent (diluent) is some type of oil. The PLS (pregnant leach solution) is mixed to emulsification with the stripped organic and allowed to separate.[citation needed] The metal will be exchanged from the PLS to the organic they are modified.[clarification needed] The resulting streams will be a loaded organic and a raffinate. When dealing with electrowinning, the loaded organic is then mixed to emulsification with a lean electrolyte and allowed to separate.

The metal will be exchanged from the organic to the electrolyte. The resulting streams will be a stripped organic and a rich electrolyte. The organic stream is recycled through the solvent extraction process while the aqueous streams cycle through leaching and electrowinning[clarification needed] processes respectively.[citation needed]

Ion exchange

Chelating agents, natural zeolite, activated carbon, resins, and liquid organics impregnated with chelating agents are all used to exchange cations or anions with the solution.[citation needed] Selectivity and recovery are a function of the reagents used and the contaminants present.

Metal chelation

Often hydrometallurgy involves the use of chelating agents, which can selectively bind certain metals. Such chelating agents are typically amines of schiff bases.

METAL RECOVERY

Metal recovery is the final step in a hydrometallurgical process. Metals suitable for sale as raw materials are often directly produced in the metal recovery step. Sometimes, however, further refining is required if ultra-high purity metals are to be produced. The primary types of metal recovery processes are electrolysis, gaseous reduction, and precipitation. For example, a major target of hydrometallurgy is copper, which is conveniently obtained by electrolysis. Cu2+ ions reduce at mild potentials, leaving behind other contaminating metals such as Fe2+ and Zn2+.

Electrolysis

Electrowinning and electrorefining respectively involve the recovery and purification of metals using electrodeposition of metals at the cathode, and either metal dissolution or a competing oxidation reaction at the anode.

Precipitation

Precipitation in hydrometallurgy involves the chemical precipitation of either metals and their compounds or of the contaminants from aqueous solutions. Precipitation will proceed when, through reagent addition, evaporation, pH change or temperature manipulation, any given species exceeds its limit of solubility.

LEACHING (CHEMISTRY)

Leaching is the process of extracting minerals from a solid by dissolving them in a liquid, either in nature or through an industrial process. In the chemical processingindustry, leaching has a variety of commercial applications, including separation of metal from ore using acid, and sugar from beets using hot water. In a typical leaching operation, the solid mixture to be separated consists of particles, inert insoluble carrier A and solute B. The solvent, C, is added to the mixture to selectively dissolve B. The overflow from the stage is free of solids and consists of only solvent C and dissolved B. The underflow consists of slurry of liquid of similar composition in the liquid overflow and solid carrier A. In an ideal leaching equilibrium stage, all the solute is dissolved by the solvent; none of the carrier is dissolved. The mass ratio of the solid to liquid in the underflow is dependent on the type of equipment used and properties of the two phases.

Leaching is the process by which inorganic, organic contaminants or radionuclides are released from the solid phase into the waterphase under the influence of mineral dissolution, desorption, complexation processes as affected by pH, redox, dissolved organic matter and (micro)biological activity. The process itself is universal, as any material exposed to contact with water will leach components from its surface or its interior depending on the porosity of the material considered.

One such reaction is:

$$Ag_2S + NaCN \rightarrow Na[Ag(CN)_2] + Na_2S$$

LEACHING PROCESSES FOR BIOLOGICAL SUBSTANCES

Many biological organic and inorganic substances occur in a mixture of different components in a solid. In order to separate the desired solute constituent or remove an undesirable solute component from the solid phase, the solid is brought into contact with a liquid. The solid and liquid are in contact and the solute or solutes can diffuse from the solid into the solvent, resulting in separation of the components originally in the solid.

This separation process is called liquid-solid leaching or simply leaching. Because in leaching the solute is being extracted from the solid this is also called extraction. In leaching, when an undesirable component is removed from a solid with water, the process is called washing. In the biological and food processing industries, many products are separated from their original natural structure by liquid-solid leaching.

An important process for example is the leaching of sugar from sugar beets with hot water. In the production of vegetable oils, organic solvents such as hexane, acetone, and/or ether are used to extract oil from nuts beans and seeds. In the pharmaceutical industry, many different pharmaceutical products are obtained by leaching plant roots, leaves, and stems.

LEACHING PROCESSES FOR INORGANIC AND ORGANIC MATERIALS

Leaching is extensively used in metal processing industries. The useful metal may occur in mixtures with very large amounts of undesirable constituents, and leaching is used to remove the metals as soluble salts. The use of acids is prevalent in the metal processing industry, Sulphates are normally used to remove metals from the solid phase, these produce harmful environmental byproducts such as sulphates.

SHRINKING-CORE MODEL

As mentioned above when leaching a solid with a liquid the desired solid goes to the liquid phase while undesired solid remains. The removal of the solid as the liquid dissolves into the particle leads to a diameter of unleached core that shrinks with time. A mathematical model can be used for this process, derived using Fick's law of diffusion taking pore diffusion as rate limiting step:

$$t = \frac{\rho_B r_s^2}{6 D_e b M_B c_{A_b}} \left[1 - 3\left(\frac{r_c}{r_s}\right)^2 + 2\left(\frac{r_c}{r_s}\right)^3 \right]$$

ENVIRONMENTALLY FRIENDLY LEACHING®

Some recent work has been done to see if organic acids can be used to leach lithium and cobalt from spent batteries with some success. Experiments performed with varying temperatures and concentrations of malic acid show that the optimal conditions are 2.0 M/L of organic acid at a temperature of 90 °C. The reaction had an overall efficiency exceeding 90% with no harmful byproducts.

$4\ LiCoO_2$(solid) + 12 $C_4H_6O_5$(liquid) → 4 $LiC_4H_5O_5$(liquid) + 4 $Co(C_4H_6O_5)2$(liquid) + 6 H_2O(liquid) + O_2(gas)

The same analysis with citric acid showed similar results with an optimal temperature and concentration of 90 °C and 1.5 molar solution of citric acid.

AMALGAM (CHEMISTRY)

An amalgam is a substance formed by the reaction of mercury with another metal. Almost all metals can form amalgams with mercury, the notable exception being iron. Silver-mercury amalgams are important in dentistry, and gold-mercury amalgam is used in the extraction of gold from ore.

KNOWN AMALGAMS AND THEIR USES

Dental amalgam

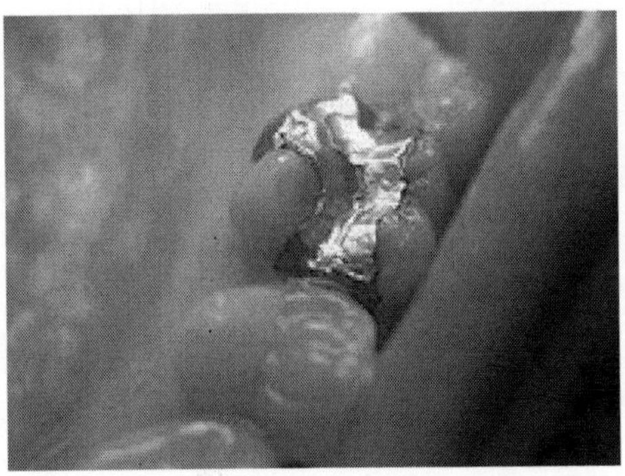

Fig. An amalgam dental filling

Dentistry has used alloys of mercury with metals such as silver, copper, indium, tin and zinc. Amalgam is an "excellent and versatile restorative material" and is used in dentistry for a number of reasons.[citation needed] It is inexpensive and relatively easy to use and manipulate during placement; it remains soft for a short time so it can be packed to fill any irregular volume, and then forms a hard compound. Amalgam possesses greater longevity when compared to other direct restorative materials, such as composite. However, this difference has decreased with continual development of composite resins.

Amalgam is typically compared to resin-based composites because many applications are similar and many physical properties and costs are comparable.

Potassium amalgam

For the alkali metals, amalgamation is exothermic, and distinct chemical forms can be identified, such as KHg and KHg2. KHg is a gold-coloured compound with a melting point of 178 °C, and KHg2 a silver-coloured compound with a melting point of 278 °C. These amalgams are very sensitive to air and water, but can be worked with under dry nitrogen. The Hg-Hg distance is around 300 picometres, Hg-K around 358 pm.

Phases K5Hg7 and KHg11 are also known; rubidium, strontium and barium undecamercurides are known and isostructural. Sodium amalgam (NaHg2) has a different structure, with the mercury atoms forming hexagonal layers, and the sodium atoms a linear chain which fits into the holes in the hexagonal layers, but the potassium atom is too large for this structure to work in KHg2.

Sodium amalgam

Sodium amalgam is produced as a by product of the chloralkali process and used as an important reducing agent in organic and inorganic chemistry. With water it decomposes into concentrated sodium hydroxide solution, hydrogen and mercury, which can then return to the chloralkali process anew. If absolutely water-free alcohol is used instead of water, an alkoxide of sodium is produced instead of the alkali solution.

Ammonium amalgam

In 1808, an ammonium amalgam was discovered by Humphry Davy and Jöns Jakob Berzelius. It is a grey, soft, spongy mass which decomposes readily at room temperature or in contact with water or alcohol:

$$2\ H_3N\text{-}Hg\text{-}H \xrightarrow{\Delta T} 2\ NH_3 + H_2 + 2\ Hg$$

It is highly toxic and harmful to the environment.

Thallium Amalgam

Thallium amalgam has a freezing point of -58 °C, lower than mercury's, and so has found a use in low temperature thermometers.

Tin Amalgam

Tin amalgam was used in the middle of the 19th century as a reflective mirror coating.

USE IN MINING

Mercury has been used in gold and silver mining because of the convenience and the ease with which mercury and the precious metals will amalgamate. In gold placer mining, in which minute specks of gold are washed from sand or gravel deposits, mercury was often used to separate the gold from other heavy minerals.

After all of the practical metal had been taken out from the ore, the mercury was dispensed down a long copper trough, which formed a thin coating of mercury on the exterior. The waste ore was then transferred down the trough, and gold in the waste amalgamated with the mercury. This coating would then be scraped off and refined by evaporation to get rid of the mercury, leaving behind somewhat high purity gold.

Mercury amalgamation was first used on silver ores with the development of the patio process in Mexico in 1557. There were also additional amalgamation processes that were created for processing silver ores, including pan amalgamation and the Washoe process.

Gold amalgam

Produced when mercury is used for gold extraction.

Gold Extraction (mining)

This has proved effective where gold fines ("flour gold") would not be extractable from ore using hydro-mechanical methods. Large amounts of mercury were used in placer mining, where deposits composed largely of decomposed granite slurry were separated in long runs of "riffle boxes", with mercury dumped in at the head of the run.

The amalgam formed is a heavy solid mass of dull gray color. (The use of mercury in 19th century placer mining in California, now prohibited, has caused extensive pollution problems in riverine and estuarine environments, ongoing to this day.) Sometimes substantial slugs of amalgam are found in downstream river and creek bottoms by amateurwet-suited miners seeking gold nuggets with the aid of an engine-powered water vacuum mounted on a float.

Gold extraction (ore processing)

Fig. Interior of the Deadwood Terra GoldStamp Mill. Crushed ore is washed over mercury coated copper sheets, whence fine gold particles form an amalgam with the mercury. The amalgam was scraped off and the gold then separated from the amalgam by heating and evaporating the mercury which was then recovered by acondenser for reapplication to the plates

Where stamp mills were used to crush gold-bearing ore to fines, a part of the extraction process involved the use of mercury-wetted copper plates, over which the crushed fines were washed. A periodic scraping and re-mercurizing of the plate resulted in amalgam for further processing.

Gold extraction (retorting)

Amalgam obtained by either process was then heated in a distillation retort, recovering the mercury for reuse and leaving behind the gold. As this released mercury vapors to the atmosphere, the process could induce adverse health effects and long term pollution. Today, mercury amalgamation has been replaced by other methods to recuperate gold and silver from ore in developed nations. Hazards of mercurial toxic waste have played a major role in the phasing out of the mercury amalgamation processes. However mercury amalgamation is still regularly used by small-scale gold placer miners (often illegally) & particularly in developing countries.

AMALGAM PROBE

Mercury salts are, compared to mercury metal and amalgam, highly toxic due to their solubility in water. The presence of these salts in water can be detected with a probe that uses the readiness of mercury ions to form an amalgam with copper. A nitric acidsolution of salts under investigation is applied to a piece of copper foil, and any mercury ions present will leave spots of silvery-coloured amalgam. Silver ions leave similar spots but are easily washed away, making this a means of distinguishing silver from mercury.

The redox reaction involved where mercury oxidizes the copper is:

$$Hg2+ + Cu \rightarrow Hg + Cu2+$$

6

Land Surveying

INTRODUCTION

Land surveying (or cadastral surveying as it is more properly known) is the definition of land boundaries when land is subdivided, bought or sold. Cadastral surveying is one discipline within a survey industry that encompasses mining, engineering and hydrographic surveying, geographic information systems, photogrammetry and geodesy. Land surveying and mine surveying are the only surveying disciplines that are directly regulated by the Western Australian government (and by State governments in other jurisdictions). Regulation of mine surveying is considered necessary for the safety and security of operations in open-pit and underground mines. Regulation of land surveying is considered necessary to protect both private interests (surveyors' clients and third parties) and the public interest (an accurate cadastre to support orderly land administration).

Each State has its own system of regulating land surveying, which typically involve:

- Occupational registration and/or licensing; and
- Controls over survey practice and standards (through regulation and enforcement).

OCCUPATIONAL LICENSING

To practice as a land surveyor in Western Australia a person is required by the Licensed Surveyors Act 1909 to hold a licence and a current practicing certificate. There are 650 licensed surveyors in Western Australia, of whom 250 hold a current practicing certificate. Various criteria must be met to obtain a licence, including obtaining a certificate of competency (S7, 8,11A), for which the surveyor must complete four years professional training and two years practical experience. A surveyor holds a licence for life (unless it is withdrawn through disciplinary action or voluntarily surrendered). Authorised surveys are the bread-and-butter of licensed surveyors, and to carry these out the surveyor must hold a current practicing certificate (S16). Practicing certificates may be obtained for a period of up to three years, but the surveyor may be required to undertake ongoing training to maintain the certificate during that period and to renew it upon expiry (S11B). Licensed surveyors are entitled to undertake authorised surveys of mining tenements under the Mining Act 1978, but must first obtain the status of "approved surveyor" from the Act's responsible Minister. Approximately 85 firms and individuals are "approved surveyors".

CONTROL OF SURVEY PRACTICE

Authorised surveys set out the boundaries of a parcel of land. Along with the registration of interests on title, the survey is one of the foundations of land ownership in Australia. Authorised surveys have legal status as evidence of boundaries when the plan is lodged with, and approved by, a relevant government agency. State governments each have a system of acts, regulations and guidelines for controlling the practice of authorised surveys. These generally specify the technical standards and procedures to ensure the accuracy, consistency and clarity of the survey and resulting plans. Invariably, such legislation contains enforcement provisions and penalties to ensure that surveyors comply with procedures and standards. In Western Australia, the rules for authorised surveys are set out in the Licensed Surveyors (Guidance of Surveyors) Regulations 1961 and supporting guidelines, and in other Acts and regulations affecting specific types of land title.

The most important of these are:

- The Transfer of Land Act (which administers freehold lands)
- The Strata Titles Act (which administers strata title land)
- The Land Administration Act (which administers Crown land)
- The Mining Act (which administers mining tenements)

OVERVIEW OF LICENSING BOARD

LEGAL STATUS

The Land Surveyors Licensing Board is a statutory body established through the Licensed Surveyors Act 1909. This is an Act to:Provide for the Registration and Licensing of Land Surveyors; to control the Practice of such Surveyors, and for other purposes incidental thereto. The Board can sue or be sued (S24).

Location

The Licensed Surveyors Act, and accordingly the Board, have traditionally been under the administration of the Minister for Lands. They was moved into the Planning and Infrastructure portfolio in 2001, but as of 1 July 2003 were transferred to the new portfolio of Minister for Land Information (the Hon. Nick Griffiths MLC). The Board has historically had a close association with the Department of Lands and Survey (later DOLA, now DLI). This relationship continues today, with DLI providing free accommodation for the Board's administrative office, and subsidising the Board's staff and operations.

Funding

The Board estimates that its operating cost is in the order of $140,000 per annum. The Board generates around $40,000 from fees for licensing and related services. DLI provides around $100,000 towards the cost of the Board. This sum consists mainly of the salary of a DLI officer who serves as the Board's secretary, and notional charging for accommodation and office services. The Board has its own bank account and its monies are not part of government's Consolidated Fund.

Composition

Table: Membership of Land Surveyors Licensing Board Current Arrangements

	Current Arrangements		Future Arrangements (under bill)	
	Member	**Required criteria**	**Member**	**Required criteria**
1	Surveyor General (SG)	Professional officer charged with the general control of surveys in department	SG	The requirements of the office of SG are under review.
2	Nominee of SG	Nil (but appointee selected for diversity and consumer interests)	Nominee of Minister	To represent users of surveyor's services
3	Nominee of SG	Licensed surveyor	Nominee of DLI CEO	To represent interests in relation to land registration matters
4	Nominee of WA Institution of Surveyors	Licensed surveyor	Nominee of WA Institution of Surveyors	Licensed surveyor and holder of a practicing certificate
5	Nominee of WA Institution of Surveyors	Licensed surveyor	Nominee of WA Institution of Surveyors	Licensed surveyor and holder of a practicing certificate
6	Nominee of academia	Member of teaching staff of a surveying course	Nominee of academia	Member of teaching staff of a surveying course
Chair	Surveyor General (ex officio)		Nominated by Minister and appointed by Governor from the six members	

The Act requires that the Board comprise six persons who represent government, professional and academic interests in cadastral surveying, including the Surveyor General, who is an ex officio member and chairman.The membership and chairmanship of the Board are likely to change as a result of the Acts Amendment and Repeal (Competition Policy) Bill 2002. The proposed changes in the Bill are set out in the table below.

Charter and Objectives

Although not required by the Act, the Board has developed a Charter and Objectives statement to guide its operations. This document was approved by the Minister for Lands in 1987. The Charter states that the Board is constituted: To provide for the registration, regulation and supervision of persons who undertake surveys and to assist the profession in the control of matters incidental thereto.

Under the Charter, the Board's principal objectives are to:

- Promote the provision of sound education requirements for the profession coupled with continuing assessment;
- Maintain reciprocity between the States and Territories of Australia and certain other countries and provide a licensing or registration structure for those persons involved in authorised surveys;
- Provide a means of administering discipline in cases of breaches of the Act in accordance with the relevant sections of the Act ;
- Ensure sound liaison with Government; and
- Ensure sound liaison with Industry.

Statutory Functions

While the Board operates under its Charter and Objectives, its primary authority derives from the Licensed Surveyors Act. The Board has acknowledged this, referring for example, to 'its role as defined by the Licensed Surveyors Act.' In its description of the functions and powers of the Board, the Act is not as precise and clear as comparable modern legislation. The Board's functions are scattered throughout the Act and are often described incidentally or through inference.

Close scrutiny suggests that the Act confers the following functions on the Board:

- To issue a certificate of competency required to obtain a surveyors licence (S9)
- To issue licenses under the Act (S7, 11)
- To issue and renew practicing certificates (S11A)
- To maintain a public register of licensed surveyors (S12, 13 and 14)

Statutory Powers

The Act assigns various powers to the Board. For ease of explanation, these have been grouped as follows.

Administrative Powers:

- To appoint officers (S6)
- To apply funds to Board's operations (S25)
- To make regulations (S26 and 26A)

Licensing Powers:

- To enter into reciprocal agreements to recognise licenses granted within other States and countries (S10)
- To require licensed surveyors holding practicing certificates to undergo continuing professional development (CPD), to charge fees for such courses and services, and to take measures against those who do not meet CPD requirements (S11B)

Inquiry and Enforcement Powers:

- To require people to attend and answer questions (S8)
- To hold inquiries into written charges of misconduct and negligence made against licensed surveyors (S21 and 22)
- To impose licensing and financial penalties on licensed surveyors that it has found to be guilty of misconduct and/or negligence (S21)
- To institute legal actions against others and defend itself in proceedings (S24)

POWER TO MAKE REGULATIONS

The Board's regulation powers warrant further description. Section 26 states that: The Board may, with the approval of the Governor, make regulations under the Act prescribing all matters that are required or permitted by this Act to be prescribed, or are necessary or convenient to be prescribed for giving effect to the purposes of this Act ...

Sections 26 and 26A specify matters for which the Board may make regulations. For ease of explanation, these have been grouped into the following areas:

Administrative and Procedural Regulations

- The administration of the Board, the register, and associated matters (S26 (a) – (d))
- The issue of certificates of competency, licences, practicing certificates and long-service certificates (S26 (g), (ia))
- The registration of contracts and articles of apprenticeship (S26 (k))
- The prescription of a scale of fees for administrative services, examination services, and Board meeting attendance (S26 (j))
- The conduct of proceedings in connection with charges of misconduct (S26 (l))

Licensing Qualifications Regulations

- Evidence to be produced and conditions to be fulfilled by an applicant for a licence or practicing certificate (S26 (e))
- Education and training requirements for a certificate of competency (S26 (f))

- The types of examinations and qualifications that will be recognised by the Board as a qualification for the issue of a licence (S26 (h) (i))

Professional Indemnity Regulations

- The Board may make regulations for various professional indemnity matters under S26A (but after consideration has chosen not to make such regulations at this stage).

Survey Practice Regulations

- The Board may make regulations for 'the direction and guidance of licensed surveyors in carrying out provisions of the Acts relating to authorised surveys' (S26 (m)).

Under these provisions, the Board has made the following regulations:

- Licensed Surveyors (Licensing and Registration) Regulations 1990

These regulations deal with arrangements and procedures for the licensing and registration of surveyors, including fees for Board services.

- Licensed Surveyors (Guidance of Surveyors) Regulations 1961

These regulations deal with the practice of authorised surveys and apply to surveyors licensed under the Act (Regulation 3).

- Licensed Surveyors (Transfer of Land Act 1893) Regulations 1961

The regulations deal with survey practice under the Transfer of Land Act.

Non-Statutory Functions and Roles

The Board undertakes a range of functions and roles that are not explicitly required by the Act. However, these functions are either necessary or desirable for the Board to perform its statutory functions, including:

- Monitoring of industry and technology developments
- Liaison with DLI regarding survey standards and performance
- Liaison with industry
- Providing advice to the Minister (occasional function)

Accountability

The Act requires the Board to:

- Keep proper accounts, prepare financial statements and obtain an independent financial audit (S25A, 25B);
- Furnish an annual report of proceedings to the Minister (S25C) ; and
- Follow rules set out in Schedule 2 as to its business, conduct and proceedings (S5).

Minister's Role and Functions

The Act confers the following responsibilities on the Minister:

- Approve fees charged by the Board for courses and services (S11B)
- Approve an auditor for the Board's accounts and financial statements (S25B)

- Receive from the Board an annual report of its proceedings, a copy of its financial statements and auditor's report, and table these in Parliament (S25C)

The Minister has no other functions or powers under the Act, such as a power to provide directions to the board or obtain information from it.

Under the normal administrative process of government, the Minister would approve and submit to the Governor in Executive Council any regulations that the Board proposed to make.

INDUSTRY-SPECIFIC FEATURES

To assist in addressing the DPC's review criteria, and to adequately consider future arrangements for the Board, it is useful to first to consider three factors that are particular to the land surveying profession. These are:

- The State's interest in the licensing of land surveyors;
- The State's interest in regulating the practice of cadastral surveying; and
- The relationship between the State and the licensed surveyor.

It is worth noting that the Board considers these factors vital in understanding its role.

STATE'S INTEREST IN LICENSING OF SURVEYORS

Government control of entry to the land surveying profession is widely accepted throughout Australia and New Zealand, and is considered necessary to manage a series of risks that are inherent in cadastral surveying. These risks are that:

- Survey inaccuracies can result in loss to the client of a surveyor;
- Survey inaccuracies can result in loss to adjoining landowners;
- Survey inaccuracies can result in loss to future landowners;
- Redress through litigation is likely to be costly and may be ineffective (the potential time lag in detecting errors may mean that there is nobody to sue);
- Loss of confidence in the accuracy of the cadastre may adversely affect the land market and, in turn, the general economy;
- Loss of confidence in the accuracy of the cadastre may compromise a wide range of government outcomes in land administration and resource management; and
- Survey inaccuracies can result in claims against the State.

The first four of these points deal with protecting a 'private good' while the latter three deal with protecting the 'public good'. In this respect, the State's interest in licensing cadastral surveyors is broader than its interest in licensing other occupations, where the focus is chiefly on the protection of individual consumers. The Board believes that its responsibility as the licensing agency is to protect both the private and the public good. Significantly, the Board sees the public good as of the highest importance. In its submission to the Minister on the Future of DOLA Report (2002), the Board stated that: The primary role of the Board is to protect the public interest in the definition of land boundaries. The Board also argued that: The Board is established under the Licensed Surveyors Act to protect the public interest in the cadastre. This protection is not only for the immediate consumer of surveying services but more importantly the government interest in a stable land system and the rights of third parties (neighbours, future owners etc) affected by the work of surveyors.

State's Interest in Controlling Practice and Standards

The State's interest in how cadastral surveyors carry out their profession is managed through prescriptive regulations made under the Licensed Surveyors Act and regulations made under other acts affecting title to land. In broad terms, the regulations deal with the following aspects of a surveyor's work:

- Calibration of equipment and methods of measurement;
- Methods of field work and boundary marking;
- The recording of field work and drawing of plans and diagrams; and
- Procedures for lodging survey information with government agencies.

These rules are designed to ensure that surveys are conducted in an accurate and consistent fashion, in order to protect the interests of individual and adjoining landowners, and the integrity of the cadastre. Control of the practice and standards of authorised surveys has traditionally rested with the LSLB. The Board introduced survey practice regulations in 1910 and 1911. The core of these regulations has endured with minimal amendment to today, and is found in:

- The Licensed Surveyors (Guidance of Surveyors) Regulations 1961; and
- The Licensed Surveyors (Transfer of Land Act 1893) Regulations 1961.

RELATIONSHIP BETWEEN STATE AND LICENSED SURVEYOR

The relationship between State and surveyor is often seen as critical to the role of the Board. The Board, on behalf of the State, grants the surveyor a licence and practicing certificate to operate in a market where competition is restricted. The State requires that the surveyor observe certain duties and obligations as a condition of maintaining the licence and certificate.

A key aspect of this is the surveyor's duty to consider the interests of third parties, the public and the government. These requirements - while not stated in the Act – are clearly set out in supporting regulations made by the Board. For example, a surveyor is required to:

(a) To work in a professional manner and to study the interests of the State in all his or her operations;
(b) To disclose all doubts, discrepancies and difficulties; and
(c) To afford to the Surveyor General all information obtained by him or her in the due performance of surveys entrusted to him or her.

A separate regulation, for surveys which are made under the Transfer of Land Act, is more explicit and bears the heading, 'Duty to the Government': The surveyor will be expected to disclose all doubts, discrepancies, and difficulties, and to afford all other information obtainable by him relative to the property, that may aid in securing accuracy and completeness in the certificate of title to the land. In these matters he must consider himself rather an agent and adviser of the Government than of the person employing him, nor will a regard for the interests of such employer be considered as excusing in any degree the withholding of any information affecting the merits of the application or dealing, even though the description supplied be literally and technically correct. This regulation is widely understood to mean that the surveyor is required to act for both government and client, and to act for the government first and foremost. The Board has argued that this, in effect, makes the surveyor a 'public officer': By registering surveyors to carry out cadastral surveys the Board is in practice making them public officers. DOLA has also argued that 'the surveyor is considered to be firstly a Public Officer'. However, it should be noted that the term 'public officer' lacks a clear or formal definition, and while it carries a sense of official status, it is not is found anywhere within the Licensed Surveyors Act or its regulations, nor is it widely used elsewhere within government.

RELEVANCE AND INDEPENDENCE

The review criteria set out by the Department of the Premier and Cabinet are wide-ranging, but essentially require consideration as to whether:

- The functions undertaken by the LSLB are essential to government objectives;
- There is a continuing need for the functions to be undertaken by a board; and,
- The functions could be undertaken more effectively by another arm of government or by the non-government sector.

These points will be considered in relation to the Board's principal areas of activity:

- the occupational licensing of the land surveying profession
- the control of practice and standards of land surveying

OCCUPATIONAL LICENSING: RELEVANCE TO GOVERNMENT OBJECTIVES

The function of licensing cadastral surveyors supports government objectives in the following areas.

- Land administration (and public administration)
- Consumer protection
- Economic development (through the operations of the land market)

As outlined earlier, occupational licensing is an essential measure to manage risks that are inherent in cadastral surveying, and which may adversely affect consumers of survey services, as well as the community and the government. The State government endorsed the ongoing licensing of surveyors as a policy position as recently as 1999, following a national competition policy (NCP) review. The policy position was supported by DOLA, the Department of Treasury and Finance and the State Cabinet.

The NCP review found that the licensing of surveyors 'is justified as being in the public interest.' It also found that the requirement for licensees to hold a current practicing certificate has resulted in estimated savings of $1,000,000 for government, by shifting primary responsibility for the verification of survey accuracy from government (DOLA) to the survey professional. Governments in all other Australian States and territories, and in New Zealand, have undertaken similar reviews in recent years, and elected to continue the registration and/or licensing of land surveyors. As a result, it is considered desirable and essential for the State to continue to regulate entry to the land surveying profession through the licensing system.

ROLE OF BOARD IN DELIVERY OF LICENSING FUNCTION

The discussion now turns to whether an independent board is required to undertake the licensing of surveyors on behalf of the State. The discussion is based on the understanding that the Board's function of administering discipline for the breach of licence conditions (including the breaches of practice and standards) is to transfer to the State Administrative Tribunal. The Board is expected, however, to gain additional investigatory powers and take cases to the Tribunal for judgement.

Relevance

There are considerable benefits accrued from having a board administer the licensing of surveyors. The most important of these is that a board provides representation for the range of groups who have a direct interest in the licensing of cadastral surveyors: government, industry, vocational training institutions, and consumers of survey services. More importantly, the Board gives these groups a management role in the licensing of surveyors. This means that affected groups each have a voice in the decision-making process as to the standards required to gain entry to the land surveying profession.

Effectiveness

A board structure brings together a wide range of expertise and experience to operate and administer the licensing system. For example, academic representation can provide input to the continuing education requirements of the licensing system; industry members bring detailed knowledge of survey practice; while government members can represent the needs and requirements of government.

The breadth and level of expertise provided by a representative board also supports the function of investigating breaches of licence conditions by surveyors, providing sound knowledge of government requirements, industry practice and surveyors' skills. It is doubtful whether the benefits identified above could be obtained more effectively, or efficiently, through a government or industry-based system of licensing. One measure of the effectiveness of the Board model is the low level of complaints made about licensed surveyors. There are on average less than 10 complaints per annum and many of these are of a minor nature. The Board has an average turnaround time of eight weeks from the receipt of a complaint to the issuing of a written finding and ruling.

Independence

While there are operational benefits arising from a representative board, there are few compelling reasons to retain a board for reasons of political independence. On behalf of government, the Board implements a public policy to license land surveyors. There would be no requirement or rationale for it to develop or exercise policy independently.

At the same time, there is a low risk of political interference in the licensing system (where a Minister might seek to influence a decision in favour of, or against, an applicant for a licence). The risks are considered low because the market is highly competitive and not particularly lucrative, and the profession is small and well known to industry associations. Furthermore, decisions to grant or withhold a licence would be subject to appeal in the State Administrative Tribunal. The Board's independence (with a mix of government and industry representatives) does minimise a potential conflict of interest that might arise if the profession alone controlled entry to the land surveying market. On the other hand, some surveyors believe the Board lacks sufficient independence from government and DOLA.

Other Models – Department

The transfer of the surveyors licensing function from an independent board to a department would be consistent with the machinery of government reforms. These reforms seek to incorporate the functions of statutory authorities (and board and committees) within departments, where appropriate and feasible. One other State, Tasmania, has taken this path and dispensed with its long-standing Surveyors Board. The registration of surveyors in that State is now the statutory responsibility of the Surveyor

General, with administration of the system handled by staff of the Office of the Surveyor General (a branch of the Department of Primary Industries, Water and Environment). The Director of Consumer Affairs has statutory responsibility for complaints, investigations and disciplinary action against surveyors, while the Institution of Surveyors Australia has responsibility for determining the competency of surveyors prior to registration.

However, there would be drawbacks to locating the licensing function with a department in Western Australia. Direct representation of industry and other interests would be lost, and there is a risk that the licensing system might lose credibility if administered solely by a department. Industry would most likely be suspicious and sceptical of the ability of a department to be the licensing agency, given that the LSLB itself has long-battled perceptions that it is a 'tool' of DOLA.

Moreover, the transfer of occupational licensing functions from boards to departments would not appear to be the general policy position of the government. For example, the Department of Consumer and Employment Protection (DOCEP) has advised that those licensing functions currently vested with statutory boards within the portfolio are likely to stay with boards. (However, the department is likely to directly undertake licensing functions currently vested with agencies that are not statutory boards, such as the courts and the Commercial Tribunal).

There would also be practical obstacles to overcome in transferring the licensing of surveyors to a department. DLI might have offered a home for the function but the department is to be disbanded and its functions transferred to a statutory authority with commercial powers. It would be inconsistent to disband the LSLB and assign its licensing function to DOCEP when other licensing boards in the portfolio are to be retained. A further alternative would be to transfer the licensing function to the Department for Planning and Infrastructure (DPI). However, this would add another regulatory responsibility to an already large and complex department.

Other Models – Industry

A further option would be the transfer of licensing function to the survey profession. This approach is found in South Australia, where the Survey Act 1992 provides for the registration and licensing of surveyors to be carried out by the Institution of Surveyors (SA Division). The Institution is also responsible for investigating complaints for unprofessional conduct. However, there would be barriers to adopting this model in Western Australia. In WA the survey industry is relatively small, and is undergoing a period of significant change in its strategic direction, its membership composition and the structure of its professional associations. Industry would require significant funding and support to undertake a licensing role. It is unclear how that funding would be sourced. One option would be to increase fees charged by the Board for licensing and other services, but the level required may be unsustainable, especially for smaller operators. Alternatively, government could provide direct funding to industry to operate the licensing system, or impose a levy on the registration of survey plans that could be passed onto the public. These options would require a rigorous examination of costs and viability that falls beyond the scope of this report.

Industry self-regulation may also be seen as inconsistent with national competition policy, given that one of the recommendations of the competition review of the Board was to reduce the influence of licensed surveyors in decision-making about entry to the market. There is also a risk that transferring the licensing function to industry might reduce the input of government, and weaken government's involvement in setting standards for entry to the profession, which is an essential method of protecting the integrity of the cadastre. There is also a risk that an industry association would place less emphasis on consumer protection than a more broadly representative board.

SURVEY PRACTICE: RELEVANCE TO GOVERNMENT OBJECTIVES

The control of the practice and standards of authorised surveys is the other primary way in which the State manages the risks in cadastral surveying. This function supports government objectives in the following areas:

- Consumer protection
- Land administration (and public administration)
- State economic development (through the operations of the land market)

There is considerable value in the State continuing to control survey standards and practices to protect the cadastre and consumers, and significant risks to government and the community were the State to abandon this function. Control of survey practice and standards ensures the different types of authorised surveys (Crown, freehold, strata and mining tenements) are conducted in a consistent and clear fashion, and to common standards. This is vital to the integrity of the land tenure cadastre, and also helps ensure ongoing compatibility between mining and land tenure cadastral information.

Consistent standards also assist the industry, enabling surveyors to confidently rely on the accuracy of adjoining surveys and survey marks and avoid duplication of effort. Consistent practices also serve to align survey documentation to the operational requirements of the State's land registration system, thereby producing greater effectiveness and efficiency for the department and consumers of its services.

The State controls the practices and standards of authorised surveys through regulations made under the Licensed Surveyors Act and the Transfer of Land Act. The former regulations provide an overarching framework for the control of practice and standards for all authorised surveys. These regulations are made by Land Surveyors Licensing Board (with the approval of the Governor), with research and development support from DLI staff. In addition, regulations made under other acts may impose practices and standards for a specific type of authorised survey (Crown, strata and freehold land and mining tenements). These regulations may be made variously by the Governor, a Minister or a statutory office (see Appendix 3 for details).

ROLE OF BOARD IN DELIVERY OF SURVEY PRACTICE FUNCTION

The review criteria require an assessment of the 'continuing need' for an independent board to undertake the function of controlling survey practice and standards.

Relevance

The LSLB has traditionally been the primary mechanism through which the State has controlled the practice and standards of authorised surveys. The reasons for this are historical and structural in part. When the Board was established in 1909 the survey industry was immature and heavily under the control of the government.

Furthermore, the Act prescribed that the Surveyor General was the Board's ex officio chairman. At that time, the Surveyor General was one of the most senior government officials, and was also the recognised head of the survey profession. In addition, many of the State's surveyors were government employees and under the direct control of the Surveyor General. As a result, the Board was a convenient and suitable instrument to undertake the management of survey practice and standards. However, the relevance of the Board to still undertake this function almost a century later must be questioned. The Board is no longer as closely allied to government as it once was. Government has only two representatives on the Board, and the chairmanship will no longer automatically rest with the Surveyor General.

At the same time, the LSLB plays no role in the daily administration and management of the land registration system and the cadastre. Nor does the Surveyor General have sole or direct responsibility for the registration of survey documentation and the setting of strategic priorities for land information and the development of the cadastre. Management responsibility for these functions has been dispersed across a range of areas within DLI.

Nevertheless, the LSLB retains responsibility for many of the regulations dealing with the lodgement and registration of survey documentation. This includes, for example, regulations prescribing the manner in which surveyors are to present survey information for registration, including the type of pens and inks to be used for field books, plans and drawings, and the type of information to be recorded by the surveyor. These are matters for which the Board has no operational responsibility.

These issues were considered by the Functional Review Committee that examined the roles of the LSLB, the Surveyor General, and the then Department of Lands and Surveys in 1985. The Committee argued that it was no longer appropriate to maintain regulations for the guidance of surveyors under the Licensed Surveyors Act. It argued that survey practice should instead be addressed within regulations made under the Transfer of Land Act and the then Land Act. The Committee also recommended: To maintain the integrity of the State's cadastral system the control of survey standards and practices be administered through the registration process of authorised surveys.

Independence

The Board has stated that its primary role is to 'to protect the public interest in the definition of land boundaries'. The Board believes that this requires it to perform more than just the licensing of the profession: The Board believes it has a role in ensuring that the integrity of the infrastructure supporting the cadastre is both maintained and expanded to support future development in the State. The regulating and licensing of surveyors safeguards this asset. ... The Board believes it has an ongoing role in the regulation of surveying, not just surveyors.

It is questionable whether the power to make regulations concerning the cadastre should sit with an independent board rather than with a government agency. As the Board itself has noted, the cadastre, 'is an asset of the State'. Final responsibility for the management of the cadastre rests with the relevant government agency that administers it – not with the Board. The accountability of the agency is arguably lessened if it does not have the full capacity to set standards and practices for the survey work that makes up the cadastre. At the same time, the Board cannot be held fully accountable for the accuracy and integrity of a cadastre that it does not own, manage or administer.

Effectiveness

On the other hand, it is clear that the Board has proved an effective structure for regulating survey practice and standards. A key part of this is that a board, with its mix of industry, government and academic representation, is effective in ensuring that regulations concerning survey practice are workable, reasonable and well informed. In this respect, a board provides a 'de facto' forum for consultation and negotiation between industry and government, and affords industry some control over the regulations which govern its operations.

A board is also well placed to ensure that survey practice in WA is consistent with that in other States, through its role in reciprocal licensing. Connected to this is the effective role that the Board has played in disciplining surveyors who are in breach of the regulations. The Board has demonstrated an ability to work closely with DLI in the detection of errors and faults in survey documents, and to investigate and prosecute instances of malpractice. However, while the Board is to continue to investigate such cases, its role in determining judgements and penalties is to transfer to the State Administrative Tribunal.

A further consideration is that the Board believes that its role in setting survey standards is essential to its licensing function: 'If the Board is to continue assessing the technical competencies of surveyors it must have a role in setting the technical standards.'

Other Models

The LSLB is one of a small number of occupational licensing boards (in the non-health sector) to have a statutory power to make regulations. Within many other occupational licensing acts, the ability to make regulations generally rests solely with the Governor rather than the licensing board. (The Governor is usually advised by the responsible Minister, who in turn is advised by a portfolio agency).

Where occupational licensing boards do have a power to make regulations, this is generally confined to matters such as:

- The administration of the board;
- The administration of the registration/licensing system; and
- The professional conduct of licensees.

Few licensing boards have the power to make regulations for industry practice, and those that do are mainly concerned with controlling practice to ensure public safety. It is also worth examining the powers of a licensing board closely akin to the LSLB. The Mines Survey Board (MSB) is established under the Mines Safety and Inspection Act 1994 to approve surveyors to undertake surveys of quarries and underground mines.

The statutory functions of the MSB (S82) are closely comparable to the licensing role of the LSLB. The MSB is required to:

- Examine the qualifications, experience and character of persons applying for authorized mine surveyor's certificates and issue such certificates where appropriate; and
- Deal with complaints concerning the holders of authorized mine surveyor's certificates and to suspend or cancel such certificates where appropriate.

Significantly, the Mines Survey Board has no role in controlling survey practice, other than a statutory function (S82) to:

- Advise the Minister on survey matters relating to mines and mining operations, including quarries and quarry operations.
- The power to make regulations for mines survey practice rests with the Governor, who may make regulations, 'dealing with the making of plans and surveys of mines.'

It could be argued that while the licensing functions of the Mines Survey Board and the Land Surveyors Licensing Board appear similar, they differ in a significant aspect. The licensing of mines surveyors is essentially designed to protect a 'private good' (safety and security within a private sector mining operation), while the licensing of land surveyors serves to protect both the 'private good' and a wider 'public good' of an accurate cadastre. In this respect, it could be argued that it is appropriate for the licensing agency for cadastral surveyors to regulate not only surveyors but surveying as well.

However, this concept is not found in other jurisdictions. Cadastral surveying licensing boards in Australia and New Zealand generally have no direct role in making regulations for survey practices and standards. Such boards generally only have a role of providing advice on survey matters to a Minister or Governor, and not all have this as a statutory function. The only exceptions to this are found in Victoria and Queensland, where land surveyors licensing boards have powers to make regulations for survey practice. However, new legislation has been proposed in both States that would remove the regulation-making function from the respective licensing board. Queensland's Parliament, for example, is currently considering the Surveyors Bill 2002, which will replace the 1977 Act.

One of the aims of the new legislation is 'to introduce more contemporary regulation of the [survey] industry' and in particular: The current power of the [Surveyors] Board to set standards for surveying has been removed. It is intended that the power will be with the chief executive [of the department] under the Surveying and Mapping Infrastructure Bill. Similarly in Victoria, the Surveyor's Board's power to make regulations under the Surveyors Act 1978 is removed from the Land Surveying Bill 2001 and assigned to the Governor. (An independent review has recently recommended that the development of regulations should rest with the Surveyor General).

TRANSFER OF RESPONSIBILITY FOR SURVEY PRACTICE

A question to consider would be how the State could control survey practice and standards if the LSLB no longer undertook this role on its behalf. Analysis of legislation suggests that sufficient mechanisms exist for government to control the survey practice and standards. Government already has powers (independent of the Board's) to make regulations for each of the principal types of authorised surveys (Crown land, freehold land, strata title and mining tenements).

However, regulations made under a single act would generally only apply to the specific type of survey covered by that act (ie. mining tenement, strata title). A key risk to consider would be the loss of coordinated standards and practice applying to all authorised surveys. The Board currently provides this through its Licensed Surveyors (Guidance of Surveyors) Regulations 1961, which provide an overarching framework for all forms of authorised surveys. This framework helps ensure consistency and accuracy within the land tenure cadastre. It also greatly simplifies the work of surveyors, in that they only have to learn and follow one main set of rules for measuring, marking and recording surveys, rather than half a dozen different systems. Furthermore, this system simplifies the connection of new surveys to existing survey networks.

There is no suggestion to scrap the high-level coordination provided by the Licensed Surveyors (Guidance of Surveyors) Regulations 1961. This discussion simply gives consideration to the transfer of this function from the Board to another branch of government. In this respect, two options would appear to exist. The first would be to use an existing mechanism that rests with the Registrar of Titles.

Under the Transfer of Land Act, the Registrar may make regulations providing: direction and guidance for licensed surveyors performing surveys authorised or required within the meaning of the definition of 'authorised survey' in Section 3 of the Licensed Surveyors Act 1909. This clause is effectively the same as the Board's regulation-making power, and of sufficient scope to cover all forms of authorised survey. There would be functional advantages for government in the Registrar assuming responsibility for direction and guidance regulations for authorised surveys. The Registrar has responsibility for the registration of interests in land, including survey documentation defining land boundaries.

This applies to titles and surveys of freehold, strata and Crown lands, which are registered within a single system under the control of the Registrar. Such surveys make up the bulk of the authorised surveys conducted in WA (the only exception being mining tenement surveys, which are discussed below). As a result, the Registrar has, arguably, the most influence and responsibility over the land tenure cadastre. In addition, under the Torrens system the Registrar is legally accountable for errors in the registration of interests in land (including some errors in land boundaries that appear on titles). These errors may result in claims for compensation being made against the State. It therefore would be reasonable for the Registrar to be able to set standards and practices for determining land boundaries.

The surveying industry has previously expressed concern as to the Registrar making regulations to cover all authorised surveys. Surveyors have argued that a Registrar would generally lack the detailed knowledge of the industry and profession to undertake such a role. Nevertheless, the Registrar would invariably be advised and guided by senior DLI officers such as the Inspector of Plans and Surveys, the officers appointed by the Governor to approve plans (under S18 of the LSA) and authorised land officers, who all are required to be licensed surveyors. However, some survey matters, such as the requirement for surveyors to connect to the geodetic network, do not relate directly to the Registrar's land titling functions but are directed at the State's land information needs and may require separate policy guidance. There is a risk that the broad consultative forum provided by the LSLB would be lost if the Registrar of Titles made regulations for the guidance and direction of surveyors. There are some factors to mitigate this risk. Firstly, the Transfer of Land Act states that the Registrar may only make such regulations 'after consultation with the Land Surveyors Licensing Board'. This would provide a check to help ensure that regulations made by the Registrar were consistent and coordinated, and workable for the government, industry and the profession.

The Registrar and senior staff of the Registration Services Branch are also involved in the Survey Industry Working Group, a formal (but non-statutory) forum for the industry and government to discuss survey matters, and this would provide another forum for consultation. Another issue to consider is the appropriateness of the Registrar, rather than the Board, making high-level direction and guidance regulations that would apply to the surveying of mining tenements. The close alignment that currently exists between the land and mining survey practice is valuable and is a result of the regulations that were instituted by the Board. There is a risk that the alignment may be compromised over time if 'direction and guidance' regulation making was passed from the Board to the Registrar.

However, this may not be a significant problem given that a sound structure for survey regulation is now in place. Preliminary discussions with the Department of Industry and Resources (DOIR) suggest that the general regulations as they stand are sufficient to meet its needs and that it has little ongoing contact with the LSLB on such matters. Moreover DOIR, through its Minister, has the facility to make recommendations to the Governor on specific regulations for tenement surveys and boundary marking, and the department may also directly make directions for the guidance of surveyors (under Regulation 117). However, if required the Registrar could be required to consult with the Department of Industry and Resources before making general regulations for the guidance and direction of surveyors.

An alternative model would be to transfer the Board's power to make regulations for the direction and guidance of surveyors entirely to the Governor. This would be consistent with the regulation-making powers found in most other occupational licensing acts and in acts controlling practice and standards for specific authorised surveys (ie. the Land Administration Act, Strata Titles Act, Mining Act). In such cases, the Governor makes regulations after receiving advice from the responsible Minister (who is in turn advised by a portfolio agency). This model would moderate the input of the Registrar of Titles and allow for broader input to the Minister. Either of the two options outlined above (transfer of function to Registrar or Governor) would require only minor amendment of the Licensed Surveyors Act.

If the Board were to retain the function of making regulations for the direction and guidance of authorised surveys, it should at least be required to formally consult with the Registrar of Titles, to improve accountability for the Board's decisions. Consideration should also be given to better coordination or integration of regulation-making responsibilities, with for example, four separate bodies having powers to make regulations for authorised surveys made under the Transfer of Land Act, and the Registrar and Board sharing identical powers to make direction and guidance regulations for all authorised surveys.

FUTURE ARRANGEMENTS FOR BOARD

POLICY FRAMEWORK

This paper also seeks to address future arrangements for the LSLB in the light of Cabinet's decision to restructure the State's land titling and information functions. These functions are to transfer from the DLI to a new statutory authority, which will have a commercial role and powers, and a board of management.

Implementation of the Cabinet decision has required a review of regulatory and policy functions within DLI. While the LSLB is an independent entity, it is a regulatory function closely associated with DLI. The discussion looks at options for the future location of the LSLB, and considers issues of efficiency and effectiveness (as set out in the DPC review criteria), which may be affected by the Board's location.

The options proposed are based on concepts identified earlier within this paper, that:

- It is in the private and public interest to continue to license cadastral surveyors;
- A representative board is the most appropriate agency to administer a licensing system;
- It is less appropriate for a licensing board to play a direct role in the control of survey practice and standards; and that
- The LSLB's disciplinary functions are to transfer to the State Administrative Tribunal (SAT) and its investigatory functions and powers expanded.

Accordingly, the options are based upon the Board undertaking the functions of:

- Licensing cadastral surveyors;
- Investigating complaints against surveyors and breaches of survey regulations; and
- Taking cases against cadastral surveyors to SAT for judgement and discipline.

Within these parameters, options for the Board's location have been identified within:

- The Consumer and Employment Protection portfolio (one option); and
- The Land Information portfolio (two options).

Land Information is a new portfolio established on 1 July 2003 as part of a wider reorganisation of ministerial responsibilities within Cabinet. Its Minister is the Hon.

Nick Griffiths MLC, who is responsible for the principal acts covering lands and surveys (excluding the Land Administration Act which has remained with the Minister for Planning and Infrastructure). Mr Griffiths portfolio includes the Licensed Surveyors Act, under which the Land Surveyors Licensing Board is established.

Consumer and Employment Protection Portfolio

This option would see the LSLB co-located with the department responsible for consumer protection, DOCEP. The majority of the State's occupational licensing bodies (outside of the health sector) are co-located with this department, including those covering the property sector (land valuers, and real estate and settlement agents).

Final arrangements for the relationship between portfolio licensing boards and DOCEP are still being established in the wake of a major review. However, it appears likely that the boards will continue to take responsibility for licensing, and that this would be their primary responsibility. Departmental officers will provide administrative support for the boards and licensing systems, and take responsibility for complaints investigations, policy and community education.

DOCEP has indicated that a number of matters would need to be addressed if the LSLB were to be transferred to its portfolio.

These would include:

- The extent of the LSLB's focus on consumer protection;
- Additional resourcing to support the Board, the licensing system and legislation;
- The impact on the department's structure and functions; and
- Gaining support from industry and stakeholders.

Other issues to consider are that:

- The investigation of complaints against surveyors requires considerable technical knowledge of surveying practice and its legal framework;
- Only around half of the complaints received by the LSLB arise from consumers (the rest come from government, or more specifically, DLI); and
- The regulatory framework for survey practice relates as much to government's land administration objectives as to consumer protection goals.

Pros and cons of a transfer to the CEP portfolio are considered in the table below.

PROS	CONS
Licensing function linked to Minister responsible for consumer protection.	Consumer protection is not the sole or primary focus of the licensing function. Licensing function not linked to Minister responsible for land information and the cadastre.
Licensing of land surveyors aligned to the agency with lead responsibility for occupational licensing.	Licensing of land surveyors not aligned with agency with lead responsibility for surveys and the cadastre.
Likely to be a greater focus on the 'consumer protection' goals than in other models.	Consumer protection is not the sole or primary focus of the licensing function. Risk of a lesser focus on the 'cadastral protection' goals of licensing surveyors than other models.
A department is generally an appropriate structure in which to locate regulatory and licensing activities.	DOCEP has little functional alignment to government and industry interests in land surveying.
Potential to improve consumer protection outcomes for the community.	Risk that government and public interest outcomes may be adversely affected.
Potential for 'economies of scale' through having one central department servicing a variety of boards.	Risk of reduced efficiency and effectiveness if the LSLB is distant from records and systems of the agency responsible for survey information. Risk that DOCEP would lack specialist resources to effectively support the Board and licensing function.

Land Information Portfolio

This is one of two options for the LSLB to remain with the Minister responsible for land titling and survey legislation.

This option would see the Board co-located with the statutory authority that is to take over DLI's land titling, land information and property valuation functions. This option is not consistent with machinery of government reforms to separate regulatory functions from service delivery.

However, the licensing of surveyors has considerable synergy with the authority's functions of land titling and land information, and particularly the inspection and authorisation of survey documents.

Locating the licensing function within the authority would raise concerns as to:

- Ensuring the continuation of current funding arrangements whereby government supports the Board to operate the licensing system;
- The impact on the authority's structure and functions; and
- Gaining support from industry and stakeholders.

PROS	CONS
Retains link between licensing function and Minister responsible for land information and the cadastre.	Licensing function not linked to Minister responsible for consumer protection (but this is not the sole or primary focus of the licensing function).
Licensing of land surveyors fully aligned with agency with lead responsibility for authorised surveys and the cadastre.	
Likely to be a greater focus on the 'cadastral protection' goals of licensing surveyors than DOCEP model.	Risk of a lesser focus on the 'consumer protection' goals of licensing surveyors than DOCEP model (but this is not the sole or primary focus of the licensing function).
	A statutory authority with commercial powers is not necessarily an appropriate structure to support regulatory and licensing functions.
Potential for greatest effectiveness and efficiency as the LSLB would be close to the records and systems of the agency responsible for survey information.	No potential economies of scale as found with DOCEP model (one department, several boards).
	Some risk that LSLB will be seen as lacking independence from statutory authority.

This is one of two options for the LSLB to remain with the Minister responsible for land and survey legislation. This option would see the LSLB remain an independent entity within the portfolio and function as a 'stand-alone' unit (not co-located with the statutory authority). The option would strengthen the functional independence of the Board and reduce industry perceptions that the Board has been too closely aligned with DOLA. The Board believes this option would improve its ability to advise and report directly to the Minister responsible for land and survey legislation.

The option is not entirely consistent with machinery of government reforms, which seek to centralise regulatory responsibilities into departments and generally streamline portfolios. Nor is it consistent with one of the objectives of this review (addressing the inefficiency across government of having 'vast numbers' of boards and committees, many of which report directly to a Minister). The 'stand-alone' model would require significant attention as to how the operations of the Board would be funded (given that many of its operating costs are currently supported by DLI on a notional charging basis).

PROS	**CONS**
Retains link between licensing function and Minister responsible for land information and the cadastre.	Licensing function not linked to Minister responsible for consumer protection (but this is not the sole or primary focus of the licensing function).
Licensing function located within portfolio responsible for cadastre and land information.	Licensing of land surveyors not fully aligned with agency with lead responsibility for authorised surveys and the cadastre.
Likely to be a greater focus on the 'cadastral protection' goals of licensing surveyors than DOCEP model.	Risk of a lesser focus on the 'consumer protection' goals of licensing surveyors than DOCEP model (but this is not the sole or primary focus of the licensing function).
	Risk of reduced efficiency and effectiveness if the LSLB is too distant from records and systems of the agency responsible for survey information (new statutory authority).
Provides greater scope to move to self-funded model in future.	Possibility of increased costs for the LSLB (and DLI).
Potential to strengthen the real and perceived independence of the LSLB.	
Improved ability to advise and report directly to the Minister responsible for survey and land legislation.	Not consistent with machinery of government reforms.

REVIEW ASSESSMENT: OTHER ISSUES

This final section seeks to identify matters that fall somewhat outside the scope of the review but may be pertinent to future arrangements for the Board and its Act.

IMPROVED LEGISLATION

Legal Status

The legal identity of the Board is noticeably weaker than comparable licensing boards. In establishing the Board, S4 of the Act simply states that, 'There shall be a Land Surveyors Licensing Board ...' Other acts typically set out the legal identity of a board more explicitly, such as the Land Valuers Licensing Act 1978:

The Board:

(a) Shall be a body corporate with perpetual succession and a common seal;

(b) Shall be the licensing and supervisory authority for the purposes of this Act; and

(c) Shall have the powers, duties and functions, conferred, imposed or prescribed by or under this Act.

A similar provision would improve the Board's legal identity and its relationship to the State.

Clarity of Role and Responsibilities

As mentioned earlier, the 1909 Act does not clearly set out the functions of the Board in the manner found in contemporary legislation. Nor does the Act contain any statement as to the strategic objectives of the Board or the licensing system that the Board administers. It may prove beneficial to the Board, to government and industry, if the Act was amended to clarify the roles and responsibilities of the LSLB. This process could be used to formalise some of the non-statutory functions undertaken by the Board.

Additional Functions and Powers

The Act is likely to be amended to strengthen the Board's investigatory powers and to enable it to present cases to SAT. Consideration could also be given to a formal power to use government staff and facilities to support the Board's operation and proposed investigatory functions.

Improved Accountability and Ministerial Control

The Act is unusual in that its accountability provisions are limited, as are the powers of the Minister. While the Board is the smallest agency within the Minister's portfolio, it is arguably the least accountable. For example, Ministerial powers to direct a board, and to obtain information from it, are commonly found in legislation for statutory authorities and boards, and are also found in legislation for occupational licensing boards. Typically these enable a Minister to give directions in writing to a board, and to obtain information from it. Legislation for the following licensing boards, for example, contain such provisions:

- Real Estate Agents and Business Agents Supervisory Board
- Settlement Agents Supervisory Board
- Builders' Registration Board
- Plumbers Licensing Board

Modernisation

The Licensed Surveyors Act is nearly a century old. It was the subject of some modernisation in 1987, particularly with the requirement that licensed surveyors be required to hold a current practicing certificate. National competition policy reforms are also likely to further modernise the Act. However, consideration should be given as to whether the Act and its regulations are due for a thorough overall, to take account of the vastly different structures and operational processes of modern government and a mature survey industry. Governments through Australia and in New Zealand have commenced or completed such a process in recent years - Western Australia is the only exception.

STRENGTHENED INDEPENDENCE

The Board has long-battled a perception that it is an instrument of DLI. If the Board is to remain co-located with DLI (and subsequently the statutory authority), steps should be taken to strengthen and reinforce the independent identity of the Board. These could include clarification of staffing, funding and housing arrangements.

Funding Arrangements

The Board's funding arrangements will need to be strengthened and made more transparent, regardless of whether it remains with DLI or is relocated elsewhere. In addition, the Board has indicated that its operating costs are likely to increase if:

- It is not co-located with the land titling and land information functions; and/ or
- It is no longer chaired by a senior government officer with detailed knowledge of the land titling functions and survey legislation.

Preliminary work by the Board suggests that it may require up to $200,000 per annum to fund its operations under these circumstances. The Board currently generates $40,000 per annum from its operations.

7

Coal Mining Geospatial Data Standards

INTRODUCTION

The Mine Improvement and New Emergency Response Act of 2006, also known as the MINER Act, was signed by President George W. Bush on June 15, 2006. This legislation was called by MHSA "the most significant mine safety legislation in 30 years, amends the Mine Safety and Health Act of 1977 and contains a number of provisions to improve safety and health in America's mines."

While the MINER Act is significant mine safety legislation, it failed to consider the role played in recent mining disasters by inaccurate paper maps. This situation represents a complete disconnect between the new legislation and a major factor leading to recent mining related disasters.

ROLE OF INACCURATE PAPER MAPS IN RECENT MINING DISASTERS

The information that follows was compiled to demonstrate the irrefutable cause-and-effect relationship between dependency on inaccurate mapping and limited survey control in recent national mining disasters. In some instances inaccurate mapping was a direct cause of an emergency. On other occasions it was a significant impediment to rescue operations. Survey control could likewise be characterized as both cause of an incident and as impeding rescue operations.

Martin County, Kentucky slurry impoundment failure

On Oct. 11, 2000 the nation's largest coal slurry spill occurred at the Martin County Coal Company in Inez, Kentucky. The EPA called the Inez spill the worst environmental catastrophe in the history of the Eastern United States. Far more extensive in damage than the widely known 1989 Exxon Valdez oil spill off the coast of Alaska, the Martin County Coal slurry spill dumped an estimated 306 million gallons of toxic sludge down 100 miles of waterways. To inaccurate mapping as the principal cause of this environmental disaster.

"The investigation of the spill shows that the protective barrier between an underground mine and the Martin County coal-waste impoundment was far thinner than regulators thought. Map information Martin County Coal Co. gave the state in seeking a permit to expand the impoundment showed a barrier of about 70 feet (21 m) between the bottom of the impoundment and the mine. However, ... the barrier was apparently less than 10 feet (3 m) thick."

Quecreek Pennsylvania mining accident

During the emergency the primary mechanism for working. In our digital information age, dependence on paper maps that have no way to be easily tested against other mapping information to check for inaccuracies impedes the rescue effort and could cause unnecessary fatalities.

Fortunately for the nine men involved, everyone remembers Quecreek because of the rescue effort was successful. The perception is that paper maps saved the day allowing rescuers to accurately locate and drill the rescue shaft used to free the trapped men. Unfortunately for workers trapped after Quecreek, most people never learned or quickly forgot why the Pennsylvania mine became inundated July 24, 2002. The United States Mine Safety and Health Administration issued a report July 24, 2006 citing faulty maps as a cause for the Quecreek Mine disaster in June 2002, trapping nine miners for 4 days.

Requirements for survey ground control in Pennsylvania prior to Quecreek were basically similar to what the authors believe exists today in most coal mining states. Underground maps are referenced to a group of ground control points typically clustered together near the face up of the mine. As a result of delays encountered in surveying the location to drill a rescue shaft during the successful Quecreek rescue, Pennsylvania passed reform legislation requiring survey control at one mile intervals along the axis of active underground mining operations. The lack of similar legislation in West Virginia impacted attempts to rescue miners trapped in the Sago mine approximately three and a half years later.

Sago Disaster

The source of the energy for ignition at Sago. Potential paths for the electricity into the sealed area were listed including ... "through the gas well casing and through the ground, or through the network of gas well lines on the surface and into the ground." While this mechanism for ignition at Sago remains unconfirmed, this potential of this being the cause of this incident underlines the importance of having all oil and gas well casings and pipelines in the vicinity of coal mining operations located with a high degree of accuracy. Likewise, oil and gas exploration and transmission workers need to know when their activities may be impacted by surface or subsurface mining activities and its associated infrastructure. With the increasing national demand for energy, it is a certainty that buffering distances between these two industries' field operations will diminish over time.

Black Castle Strip Mine Fatality

The 58-year old bulldozer operator with 15 years experience was fatally injured due to an ignition of natural gas. As the bulldozer operator was developing a drill bench, the blade of the machine contacted and ruptured a 16-inch low-pressure, high-volume natural gas line which immediately burst into flames.

Figure: February 1, 2006 Boone County, WV a front-end loader operator was killed removing material above a coal seam when his loader bucket penetrated a buried 20" high pressure natural gas transmission line. The fatal accident, the statement is made that "Vira told Moss to stay 100 feet away from the gas line. Neither Vira nor Moss knew the exact location of the gas line." Clearly lack of knowledge of the precise location of the pipe line was the primary cause of this mining fatality. It is important to note that this death had no relationship whatsoever to OSM's federal mining industry oversight partner MSHA.

Aracoma Disaster

The Office of Surface Mining (OSM) and the state and tribal coal mining regulatory authorities (SRAs) implementing the Surface Mining Control and Reclamation Act of 1977 (SMCRA) regularly make decisions and perform actions in controlling the potential environmental impacts of surface coal mining operations. The majority of these actions consider the proximity of the proposed or existing coal mining operation to potentially affected adjacent areas and resources. Historically, these regulatory actions have been supported through the use of paper maps containing geographic features describing the coal mining operation and adjacent areas of interest. Since 1988, the Technical Innovation and Professional Services (TIPS) program of OSM has promoted the use of Commercial-off-the-Shelf (COTS) computer software applications by the SRAs to model the potential impacts of coal mining operations. In recent years, these scientific software applications allowed coal mining features shown on mining operations maps to be converted (automated) to digital format for use in computer mapping applications.

The resulting digital geospatial features can now be stored in standardized geographic databases for reuse in multiple ways. For many reasons, efforts have been in progress within individual SRAs to digitally acquire and use coal mining geospatial features in many of their SMCRA business processes.

NATIONAL COORDINATION OF COAL MINING GEOSPATIAL ACTIVITIES

At the TIPS Steering Committee meeting held in St. Louis on May 3-5, 2005, the TIPS Steering Committee and OSM Director Jeff Jarrett concurred on the formation of a National Coal Mining Geospatial Committee (NCMGC). The NCMGC was established in late FY 2005 to promote the use of geospatial technology for implementing SMCRA.

The NCMGC is supported by TIPS and operates as a partnership between OSM and the states authorized to implement SMCRA. Committee members represent the geospatial technology interests of the states, tribes, and OSM offices. SMCRA organizations with representation in the NCMGC include the Interstate Mining Compact Commission (IMCC), National Association of Abandoned Mine Lands Programs (NAAMLP), and the Western Interstate Energy Board (WIEB).

The NCMGC will develop a national strategy for implementing and promoting the use of geospatial technology within the SMCRA community; provide solutions to problems affecting implementation of enterprise Geographic Information System (GIS) within OSM; and identify geospatial activities, policies, standards, and products that will increase the effectiveness and efficiencies of organizations working on SMCRA-related projects on a national scale.

The committee will assess the need for geospatial resources to support all regulatory aspects of mining and reclamation activities and the availability of geospatial information, systems, and expertise. The committee will evaluate the gap between the requirements, information holdings, system capabilities and expertise; identify critical geospatial tools and services for federal, state, and tribal managers and staff; facilitate improved interoperability and sharing of geospatial resources; and will help to coordinate federal geospatial resources and initiatives relevant to national requirements. Specifically, the committee will examine SMCRA business processes for application of geospatial technology, and facilitate sharing of geospatial technologies to support implementation of SMCRA. The committee will recommend guidance in reengineering business processes across the SMCRA community to gain efficiency and benefit from today's geospatial automation technology.

The overall objective of the committee is to ensure that scientifically sound geospatial data, spatial information products, technology applications, and services are provided in an efficient and cost effective manner to the SMCRA user community for use in minimizing risk and improving regulatory decision-making relevant to surface coal mining and reclamation operations.

NCMGC Accomplishments

The first meeting to plan work activities; determining geospatial technology development status among all SMCRA organizations; establishing a Geospatial Data Steward within each SMCRA organization; conducting the first National Meeting of SMCRA Geospatial Data Stewards; identifying geospatial technology development needs of SMCRA organizations; establishing a Coal Mining Spatial Data Standards ASTM Task Group from Geospatial Data Stewards offering to help develop voluntary standards for exchanging coal mining spatial data; conducting the first meeting of this task group at ASTM headquarters; obtaining vendor software training for a few qualified Geospatial Data Stewards in managing coal mining geospatial data in an enterprise environment; and successful first steps in a "proof of concept" effort to develop a geospatial infrastructure to exchange selected coal mining spatial datasets between two networked servers inside OSM's Wide Area Network (WAN).

These activities of the NCMGC support the goals of the President's Management Agenda in the expansion of e-Government to reduce redundancy; facilitate horizontal (cross-federal) and vertical (federal, state and local) information sharing; establish a direct relationship between IT and mission/program performance to support citizen-centered, customer-focused government; and maximize IT investments to better achieve mission outcomes.

NCMGC ACTIVITIES PLANNED IN FY 2007

The NCMGC will continue efforts to promote the use of geospatial technology to meet the business needs of SMCRA organizations. Activities anticipated in FY 2007 include a planning meeting to review and implement recommendations from the first National Meeting of SMCRA Geospatial Data Stewards; briefings to OSM management and SMCRA organizations on the accomplishments and progress of the NCMGC in its activities; holding three meetings of the Coal Mining Spatial Data Standards ASTM Task Group to develop voluntary spatial data standards for the first two coal mining spatial data sets of national interest: surface coal mining boundaries and underground coal mining boundaries; sending additional qualified personnel to attend vendor software training in managing coal mining geospatial data in an enterprise environment; continuing development work on a geospatial infrastructure to exchange selected coal mining spatial datasets among networked servers outside OSM's WAN; recruiting state regulatory programs to participate in this geospatial infrastructure; conducting an outreach program by giving presentations about NCMGC activities at national meetings of SMCRA organizations; exploring participation with the Mine Safety and Health Administration (MSHA); and establishing a planning sub-committee to begin preparations for an FY 08 National Meeting of SMCRA Geospatial Data Stewards with a theme of "Integrating GIS into SMCRA Business Processes".

Establishing Geospatial Data Stewards

During this two-day meeting attended by 44 participants, geospatial data stewards from 18 approved state regulatory programs and 7 OSM offices worked through issues affecting the coal mining spatial data interests of their respective offices.

Topics discussed during the meeting included promoting the use of geospatial data in coal mining, results of a recent geospatial technology development questionnaire distributed among SMCRA programs, establishing voluntary standards for coal mining geospatial data, working with the coal industry to exchange coal mining spatial data, establishing an electronic infrastructure to share coal mining geospatial data, and vendor software training for highly specialized SRA staff.

The geospatial data stewards participated in a live panel discussion on the first day of the meeting and a business networking session on the second day. During the business networking session, geospatial data stewards first identified needs, goals, and expectations and then established priorities. Working with the geospatial data stewards were representatives from the MSHA, Peabody Western Coal Company, Environmental Systems Research Institute (ESRI), and the American Society of Testing Materials (ASTM) International.

DEVELOPING COAL MINING SPATIAL DATA STANDARDS

At the National Meeting of SMCRA Geospatial Data Stewards, the NCMGC announced plans and requested volunteers to develop voluntary standards for the exchange of coal mining spatial data among the states, tribes, OSM offices, coal mining industry, and the public. Developing these standards will help advance the use of geospatial technology for implementing SMCRA. On August 2, initial selections from among the volunteers were made to establish the Coal Mining Spatial Data Standards ASTM Task Group. The task group consists of representatives from the state regulatory programs, OSM offices, the Mine Safety and Health Administration (MSHA), coal industry (National Mining Association), and the general public.

Daniel Kestner and Tom Galya will co-chair the task group and provide coordination and leadership. Joe Ritchey, a qualified professional facilitator knowledgeable and experienced in development of voluntary standards under American Society of Testing Materials (ASTM) guidelines, will provide technical support. Julie Maitra will provide coordination with the Federal Geographic Data Committee (FGDC) as the task group works on establishing an ASTM standard. Current members of the Coal Mining Spatial Data Standards ASTM Task Group are listed in Table 1 below.

Table. Members of the Coal Mining Spatial Data Standards ASTM Task Group.

Name	State/OSM Affiliation
Alan Wilhelm	OSM, Western Region Office
Arielle Feitelson	OSM Appalachian Region Office
Bruce Johnson	North Dakota Public Service Commission
Daniel Kestner (Co-chair)	Virginia Department of Mines, Minerals and Energy
Darrell Trent	Peabody Coal Company
Jo Gault	OSM, Knoxville Field Office
Joe Ritchey (Facilitator)	ASTM International
Kathy Rossmann	Ohio Division of Mineral Resources Management
Mike Shank	West Virginia Department of Environmental Protection
Robert Hughes	Luzerne (PA) Conservation District
Sandin Phillipson	Mine Safety and Health Administration
Tom Galya (Co-chair)	OSM, Charleston Field Office

ASTM International Headquarters near Philadelphia, PA. At this meeting, group members learned how the ASTM methodology works for developing voluntary national standards for geospatial data. The task group began the process to develop standards for the first two national coal mining spatial data sets – surface coal mining boundaries and underground coal mining boundaries – which will be assembled and periodically updated from existing data sets already in use within each of the various regulatory programs in differing formats.

The task group also began to identify and work on many issues to define the content, scope, geometry, attributes, and other characteristics of these first two national coal mining spatial data sets. When voluntary standards have been established for these first two coal mining spatial data sets, review and comment will be requested from all interested parties. As requirements are met for establishing ASTM standards for these two spatial data sets, requirements to establish FGDC standards will also be accomplished. Three additional team meetings are planned for various locations in the U.S to continue work on this task.

Meetings are planned for Costa Mesa, CA on January 30, Newport, VA in June, and Salt Lake City, UT in September. In future years, development of additional national layers of coal mining spatial data will be considered. These additional layers may include geologic sampling locations and their chemical attributes, surface- and ground-water sampling locations and their associated water quality attributes, bond release areas, abandoned mined land areas, or other data layers determined by the task group to be of national interest.

Creating a Coal Mining Spatial Data Infrastructure

The National Meeting of SMCRA Geospatial Data Stewards, OSM began internal discussions to determine how to develop a coal mining geospatial infrastructure to collect, store, manage, use, and distribute selected coal mining spatial data from among all participating SMCRA organizations. At the SRA level, two elements are critical to implementation. First, SMCRA Geospatial Data Stewards must adopt the data standards developed by the Coal Mining Spatial Data Standards ASTM Task Group and apply them to applicable coal mining spatial data used within their respective SMCRA organizations. Second, the SMCRA Geospatial Data Stewards must use a Relational Database Management System (RDBMS) such as Microsoft SQL Server or Oracle to store their data and also manage it with a spatial data management system such as ESRI's ArcSDE.

Through read-only access to selected coal mining spatial data in the SMCRA organization's ArcSDE geodatabase, an automated process from a centralized server within OSM's Wide Area Network (WAN) will download the selected spatial data, re-project the data from a local coordinate system to a national coordinate system and perform other data adjustments as may be required, and aggregate it with data from other SMCRA organizations into a single national dataset for each selected theme. By using an automated process, no resource requirements will be imposed by OSM on participating SMCRA organizations to collect, process, or upload "stale" coal mining spatial data to OSM, and resource requirements on OSM will be minimized. Coal mining spatial data uploaded through the automated process will be the highest quality, most reliable, and latest available data from the authoritative data source, the SMCRA organization creating and managing the data on a daily basis for their respective area.

In this manner, selected coal mining spatial data from all participating SMCRA organizations will be combined to form national datasets appropriate for internal use by OSM and for distribution to the public via Internet map server applications and/or data services to Geospatial One Stop (GOS), the

National Map, and other means as appropriate without additional human resource commitments by the participating SMCRA organizations. Many problems will need to be overcome during development of a national coal mining spatial data infrastructure. Potential problems include but are not limited to inability of small SRA programs to adequately participate due to resource issues; possible resistance to participation by SRA's; state and federal mandated Information Technology (IT) requirements adversely affecting infrastructure development; IT security issues; unanticipated difficulties within OSM in collecting, managing, and developing applications for national datasets; spatial data metadata documentation; stewardship of authoritative spatial data; data liability and confidentiality concerns; and funding for infrastructure development.

All of these problems appear to have manageable solutions. An initial meeting of the Coal Mining Spatial Data Infrastructure team on June 29 and teleconferences of September 5 and November 7 provided opportunities for team members to identify issues, discuss possible solutions, and develop plans to prototype a spatial data infrastructure. An implementation plan consisting of two phases was selected. In Phase 1, a "proof-of-concept" pilot project to exchange coal mining spatial data between at least two ArcSDE servers at remote locations within OSM's WAN would be attempted. In Phase 2, a similar effort would be made to exchange selected coal mining spatial data between an ArcSDE server inside OSM's WAN and one or more ArcSDE servers located outside the OSM WAN at a SRA. The Phase 1 attempt was successfully conducted on August 3. Prior to this attempt, a user account with read-only permission was established on an ArcSDE server located at an OSM office in Knoxville, TN to allow access through the Internet from another ArcSDE server located at an OSM office in Denver, CO.

In Phase 1, a GIS specialist in Denver manually executed a script written in ArcGIS Model Builder to automate the entire process of logging onto the ArcSDE servers in Knoxville and Denver from a workstation, downloading previously selected data from both servers, closing connections, reprojecting four datasets from different coordinate systems to a single national coordinate system, and merging them into a single dataset. In this test, 891 surface coal mining polygon boundaries and attributes of Tennessee were combined with 8 surface coal mining polygon boundaries and attributes representing 4 surface coal mining operations located on Indian tribal lands in Arizona and New Mexico. This entire process required less than a minute for completion.

Prior to attempting Phase 2, numerous requirements must be met. In Phase 2, GIS personnel within selected SMCRA organizations will be contacted to allow access and downloading of selected coal mining spatial data located on ArcSDE servers within state organizations operating outside OSM's WAN. OSM may be required to establish interconnection agreements with the SMCRA organizations, provide business plan documentation to OSM's Chief

Information Officer (CIO), and adequately address potential issues relative to security. An automation procedure similar to that used in Phase 1 will be employed. New technologies, such as ESRI's Data Interoperability Kit, Feature Manipulation Engine, and ArcGIS Server 9.2 will be evaluated. Phase 2 is planned for May of 2007.

Using Coal Mining Geospatial Data

Within the various SMCRA organizations, managers have struggled with demands to modernize business processes, improve efficiency, and produce higher quality work products in a budget-constrained environment with limited resources. A key factor affecting these demands is the high but hidden cost of managing paper information products such as narrative reports, tables of environmental data, and maps. Historically, SMCRA organizations have required that mining companies provide these information products with every permit action involving considerable quantities of paper at great expense in order to conduct analyses supporting regulatory decisions affecting surface coal mining operations.

Extracting features shown on coal mining operation maps with their associated attributes for storage and management in a Geographic Information System (GIS) allows access, analysis, and reuse of this data by modern software applications. This is a significant technological advancement in processing data contained in the information product most resistant to conversion to "intelligent data" – paper maps. With this development, opportunities now exist for re-engineering SMCRA business processes to employ automation to improve efficiency, and use of scientific software mapping applications to conduct better analyses resulting in higher quality work products.

In the modern SRA working environment, a spatial data infrastructure which uses data storage structures in a client-server architecture will promote acquisition of new spatial data from the mining industry in a digital format; aid in conversion of existing paper-based maps to digital format through scanning and digitizing; support the business processes of the organization through the use of scientific software to display, query, and map coal mining spatial data at desktop workstations; provide spatial data for use by mobile computing devices equipped with Global Positioning System (GPS) technology at coal mining operations; support Electronic Permitting (EP) initiatives; and establish a foundation for On-Line Analytical Processing (OLAP).

OLAP allows users to query summarized, multidimensional data; apply relevant business logic; retrieve information; and produce fast, consistent, and accurate information products that support decision-making in applications without manual data manipulation. For many reasons, SMCRA organizations will require coal mining spatial data infrastructures which support the present operational needs of their organizations and provide an adequate foundation to meet future challenges. At the SRA, these infrastructures will be used in the management of day-to-day activities related to technical review of coal mining permit applications and inspection of coal mining operations. At OSM, coal mining spatial data collected from SRA infrastructures and aggregated into national datasets will be used to help implement the nation's coal mining laws and regulations, promote better understanding of the potential impacts of surface coal mining operations, detect and identify problems, and provide new opportunities in the assignment of resources to resolve potential environmental issues.

These national datasets may consist of coal mining features such as surface coal mining boundaries, underground coal mining boundaries, coal haul roads, critical earth fill structures, sediment basins, geologic drill holes, surface- and groundwater monitoring locations, and other coal mining features determined to be of national interest. Sharing and using these datasets nationally provides access to valuable natural resource information otherwise stored in paper files within the various SRA program offices and subject to future loss. OSM intends to make these national coal mining spatial datasets available to the public through Internet mapping services, registration with Geospatial One Stop (GOS), publication on the National Map, and other media as appropriate.

Benefits to the Nation

Digital geospatial data describing past, present, and proposed coal mining operations can provide significant benefits to government agencies, business interests, and the public when planning land use activities in coal-bearing areas of the nation. These benefits will include improved regulation of active coal mining operations, more successful reclamation of abandoned mine lands, and better public policy at local, state, and national levels.

IMPROVED REGULATION OF ACTIVE COAL MINING OPERATIONS

Aggregation of SRA coal mining spatial datasets into a small collection of coal mining spatial data themes of national significance provides an opportunity for OSM to use these new data sources to better understand and more efficiently work with individual SRA's on issues of local importance. For the first time, OSM will have near real-time access to coal mining spatial data accurately describing location, boundaries, spatial extent, and important attributes of significant coal mining features of all coal mining operations in the United States.

These data can be used to enhance OSM's administrative capabilities by being able to locate, identify, and conduct reviews of permitted coal mining operations. Information products derived from this data can be used to help determine and verify on a state, regional, or national basis acres under permit and acres reclaimed in various phases of bond release: support or replace current methods of obtaining information for OSM's Annual Report; and answer other "state of the land regarding SMCRA" questions in near real-time.

Creation and use of national coal mining spatial datasets by OSM will support better cooperative efforts among government organizations working in new ways such as regional hydrologic investigations of subterranean water flow among interconnected underground coal mine workings, studies of potential mine blowout locations, acid mine drainage, and underground subsidence planning and investigation. Use of these data to support traditional "oversight inspections" conducted by OSM of state regulatory activities may result in reallocation of human resources by OSM from random inspections to focus on geographically-defined "problem areas" to help SRA's better implement SMCRA.

More Successful Reclamation of Abandoned Mine Lands

SMCRA organizations manage the reclamation of lands mined for coal and abandoned prior to passage of the Act through funding provided by OSM. When reclaimed, these lands support higher and better land uses. However, certain types of reclaimed coal mine features present limitations to future land uses. These include former coal refuse piles, mine spoil areas, slurry impoundments, and underground coal mines with a high potential for subsidence.

Reclaimed coal refuse piles, mine spoil areas, and slurry impoundments should not be built upon for many years unless special construction methods are used. During reclamation, these materials may be covered by only two to four feet of earth to provide adequate depth for plant growth. Reclaimed coal mining areas may be difficult for the public to recognize without prior training or experience. These coal mining areas will continue consolidating over time and may be structurally unstable without earth compaction methods. In home and building construction, excavations for basements and foundations may also expose toxic and acidic coal mine wastes resulting in loss of reclamation and construction damage. Structures built in these areas may also be susceptible to accumulation of CO_2 and other coal mine gases.

If basements and crawl spaces are excavated into recently mined and reclaimed mine spoil, CO_2 may fill the completed structure to concentrations which can cause injury or death to the occupants if not properly ventilated. Abandoned underground coal mines near the surface may be susceptible to subsidence, causing very serious structural damage as room-and-pillar mining areas collapse many years after abandonment. The OSM AML inventory only identifies areas where abandoned mines have already subsided, causing damage to existing structures, and not those locations in which abandoned coal mines may yet subside in the future. Prior public knowledge of the location of former coal mining areas is required to adequately plan for construction in these areas and address potential problems.

Better Public Availability of Coal Mining Spatial Data

Dissemination of coal mining spatial data to the nation provides opportunities for others outside the SMCRA community to benefit from state and federal coal mining spatial data resources. In the future, these data may help inform the public about the extent and potential impacts of existing and proposed surface and underground coal mining operations, reduce the amount of new construction on abandoned mine sites, help reduce the growth of the AML inventory by public awareness of the location of abandoned coal mines, support MSHA responses to underground coal mine emergencies, assist the USGS in their efforts to estimate remaining coal reserves (Tewalt et al., 2001), help update surface land use and geologic information for improved environmental impact assessments, and enhance the utility and relevance of the approximately 134,000 underground mine map images contained in the National Mine Map Repository managed by OSM by providing a "geographic footprint" which can be used as a spatial index to the location of the map image.

Recent legislation in Virginia may someday be perceived as the first State initiative tying mining activities to Community Right to Know. In 2006 the Commonwealth passed House Bill 1562 requiring that, for the sale of new dwellings the builder or owner (if the builder is not the owner), must disclose to the purchaser whether the builder or owner has any knowledge of (i) whether mining operations had previously been conducted on the property or (ii) the presence of abandoned mines, shafts, or pits.

FITNESS FOR WORK POLICIES IN COAL MINING

A questionnaire based survey of 55 coal mines was conducted in December 2008 to ascertain how the mines had implemented the fitness for work provisions of Sections 41 and 42 of the Coal Mining Safety and Health Regulation 2001. The regulations stipulate that a mine's safety and health management system must provide for controlling risks at the mine associated with personal fatigue, other physical or psychological impairment, the improper use of drugs and the excessive consumption of alcohol.

With respect to fatigue and other physical or psychological impairment and the improper use of drugs, mines must establish the criteria for assessment in agreement with a majority of workers at the mine. With respect to the improper use of drugs, if agreement is not reached with a majority of workers, the mine must revert to the use of Recognised Standard 07 Criteria for the assessment of drugs in coal mines, issued by Queensland Mines and Energy. Most mines have systems in place that adequately address the regulations but there is some variation in the way in which the safety and health management systems are implemented and applied. In addition though it appears that mines have consulted with workers, this is not documented in all cases, and it is not clear that the mine has gained agreement from a majority of workers in establishing criteria for assessment.

Regulatory Requirements-Coal

The Coal Mining Safety & Health Regulation 2001 requires:

- in Section 41, that a coal mine's safety and health management system (SHMS) must provide for controlling risks at the mine associated with the excessive consumption of alcohol; and
- in Section 42, requires that the SHMS system provide for controlling risks associated with personal fatigue, physical or psychological impairment; and the improper use of drugs.

In controlling risks associated with the excessive consumption of alcohol the SHMS must incorporate:

(a) An education program;
(b) An employee assistance program;
(c) The following assessments to decide a person's fitness for work—
  - Voluntary self-testing;
  - Random testing before starting work;
  - Testing the person if someone else reasonably suspects the person is under the influence of alcohol.

In controlling the risks associated with the personal fatigue the SHMS must incorporate:

(a) An education program;
(b) An employee assistance program;
(c) The maximum number of hours for a working shift;
(d) The number and length of rest breaks in a shift;
(e) The maximum number of hours to be worked in a week or roster cycle.

In controlling the risks associated with the improper use of drugs the SHMS must incorporate:

(a) An education program;
(b) An employee assistance program;
(c) An obligation of a person to notify the site senior executive for the mine of the person's current use of medication that could impair the person's ability to carry out the person's duties at the mine;
(d) An obligation of the site senior executive to keep a record of a notification given to the site senior executive under paragraph (c);
(e) The following assessments to decide a person's fitness for work—
  1. Voluntary self-testing;
  2. Random testing before starting, or during, work;

3. Testing the person if someone else reasonably suspects the person's ability to carry out the person's duties at the mine is impaired because the person is under the influence of drugs.

The SHMS must provide for protocols for other physical and psychological impairment for persons at the mine. If the fitness provisions for personal fatigue, other physical or psychological impairment or for the improper use of drugs provide for the assessment of workers, the site senior executive must establish the criteria for the assessment in agreement with a majority of workers at the mine. The criteria for assessment are those criteria used to determine whether a person is fit to work, *e.g.* the presence of drugs in their system, their ability to pass a computer response test, a doctor or occupational therapist says that they cannot work due to a physical injury.

Regulatory Requirements–mines and quarries

Under the Mining and Quarrying Safety and Health Regulation 2001 a person must not carry out operations at a mine, or enter an operating part of a mine, if the person is under the influence of alcohol, or is impaired by a drug, to the extent the alcohol or drug impairs, or could impair, the person's ability to safely carry out the person's duties at the mine. Under Section 9 Risk monitoring, (4)(b) requires that monitoring could be self-monitoring to detect effects of the hazard; and as an example suggests self-recognition of physical symptoms of heat stress or fatigue. In Section 86 "Worker's self-assessment of fitness level", each worker at the mine must periodically conduct a self-assessment of the worker's condition, including, for example, for effects of heat strain or fatigue, to decide if the worker is in a fit condition to carry out the worker's duties at the mine without creating an unacceptable level of risk. In section 89 "Work hours and rest breaks" a mine's safety and health management system must provide for controlling risk at the mine arising out of personal fatigue caused by excessive work hours or insufficient rest periods.

Formulation of Survey Tool

This survey questionnaire was formulated in eight sections, being:

1. Development of the sections 41 and 42 on-site fitness provisions.
2. The safety and health management system (SHMS) for excessive consumption of alcohol.
3. SHMS for the improper use of drugs.
4. SHMS for the control of risks associated with personal fatigue
5. SHMS for the control of risks associated with other physical or psychological impairment.
6. Contractor fitness for work issues.
7. Other issues that respondents may have in relation to sections 41 and 42 of the Regulation.
8. A self assessment of the mine's FFW SHMS.

Survey Results

The intent and structuring of the questionnaire was threefold;

- to determine compliance with the various elements of Sections 41 and 42, and
- to gain an appreciation of the systems that each mine has in place to adequately discharge their obligations with regard to 'fitness for work'; and
- to determine how the current system and regulations can be improved.

55 coal mines, which represented 95% of those sent the survey, responded to the questionnaire.

- 49 individual mine responses
- 1 response representing two mines
- 1 response representing three mines
- 3 responses for separate Coal Handling and Preparation Plants facilities
- 1 response for a dragline erection site

41 of the respondents were open cut mines (operating and under development) and 14 were underground mines (including one in the surface construction phase only). Of the coal mines in operation at the time of the survey only three are not represented in the survey responses.

Table. Self assessment of FFW management systems

Fitness attribute	Consolidated self ratings of 55 respondents				
	1	2	3	4	5
Alcohol	2 %	2 %	16 %	57 %	23 %
Drugs	2 %	2 %	16 %	62 %	18 %
Fatigue	2 %	6 %	27 %	59 %	6 %
P&P Impairment	4 %	10%	37 %	41 %	8 %

Scale:1-2 a less than adequate control of FFW risk

3 a marginal control of FFW risk

4-5 an acceptable control of FFW risk

Alcohol and Drugs

All respondents stated that they have a documented SHMS for the alcohol and drug provisions of the Regulation. Most mines have a combined drugs and alcohol SHMS. The vast majority of respondents had what appeared to be adequately developed and effective systems in place.

80% of respondents rated themselves as having a system effectiveness of 4 or higher (on a 1 to 5 scale) for both the alcohol and drug SHMS responses. A number of sites are currently reviewing their 'drug' management system to comply and align with the amendments made to Section 42 in 2007, in relation to involvement and agreement by a majority of mineworkers. The method of testing (urine or oral fluid) is also under review. For the development of the criteria for assessment for drugs, 90% of respondents stated that they had been developed in accordance with section 42(7) of the Regulation (the site senior executive must make a reasonable attempt to gain agreement from a majority of workers).

Only one respondent stated that they had utilised Recognised Standard 07 and one was considering adopting it, as their default 'drug procedure' due to the fact that they could not come to agreement under the provision of section 42(7) of the Regulation. 12% of respondents stated that they have or would adopt Recognised Standard 07 (partly or wholly) into their procedure, in particular the oral fluid testing aspect. The authors are of the opinion that this percentage would now be significantly higher as oral testing becomes more accepted / prevalent.

With regard to the alcohol procedure, 98% of respondents stated that they had developed it in accordance with section 41(3) of the Regulation, which requires mineworker consultation. However, only 89 % stated that they had documentary evidence of mineworker consultation for its development.

The survey requested input from respondents on the following testing and monitoring regimes:

(a) Voluntary self-testing: For alcohol, all respondents stated that they provided a 'self testing' facility for use by mine workers who may suspect that they are 'over the limit'.

For drugs, 85% of respondents stated that their mineworkers could self test for drugs either by providing test kits or making it known where they can be sourced (eg at the minesite or from a local Pharmacy).

There are varying 'rules' associated with self test procedures, the main one being the restriction on the number of times that a person cannot start work and return 'home', taking unpaid, annual or sick leave. The norm is two occasions in a 12 month period, after which the person is placed on the 'action step/disciplinary' procedure and in the case of drugs, not to return to site until they are 'clear'.

(b) Random testing: All respondents stated that they conducted either pre-start or on-shift random testing for alcohol and drugs in one form or another. Such methods as colour coded marbles, computer based random number generator, the nth number of visitor, and utilisation of the OSPAT system were stated to be used as random selection tools.

(c) Indirect FFW screening: 25% of respondents stated that they utilised the OSPAT or a similar system as a 'front line' fitness 'identifier' and a means of 'selection' for pre-start random D&A testing. This system can also be used as an 'indicator' for chronic fatigue and other impairments that my impact on 'safe performance'.

(d) Testing under suspicion: 96% of respondents stated that they had a 'testing under suspicion' system in place.

(e) Mandatory testing following an accident or incident: 97% of respondents stated that they had a system for mandatory testing, with just over half stating that they tested all involved persons and just under half stating that the testing was conducted with certain criteria being applied.

The criteria for assessment of excessive consumption of alcohol is based on blood alcohol concentration (BAC) as follows:

- 64% of mines – Zero effectively <0.01 g/100ml
- 28% of mines - 0.02 g/100ml
- 8% of mines - Various BAC level limits ranging from 0.01 to 0.05g/100ml
- 0.05 limit - two mines - for work in non operational areas one of these allowed mineworkers to return to work in operational areas when their BAC reduced to 'zero'

The criteria for improper use of drugs used by all mines are the cut-off or target concentrations considered indicative of drug use denoted in the Australian Standards. As at December 2008, the drug testing methodology in use was urine – 47 (85%), oral fluid - 19 (35%) and both – 12 (22%). Many mines are now moving over to oral fluid testing

Most mines adopt a three step counselling/disciplinary procedure as follows:

- Step 1: (1st positive result) ensure that they are familiar with mine's FFW procedures and an entry on their personal file.
- Step 2: (2nd positive result) counselling by supervisor/manager, external counselling, commitment to mine's FFW requirements, possibility of increased frequency of tests, possible final warning, and an entry on their personal file.
- Step 3: (3rd positive result) final written warning; or show cause; or mandatory termination.

In addition there is provision for instant termination in the case of wilful or serious misconduct, no matter at what step of the disciplinary procedure.

The number of positive results for the three month period from 1 July to 31 October 2008 were 177 out of 51177 (0.35%) for alcohol and 167 out of 19790 (0.84%) for drugs.

Personal Fatigue

With regard to the SHMS for personal fatigue, 65% of respondents rated themselves as having a system effectiveness of 4 or higher (on a 1 to 5 scale as shown in Table 1). 96% of respondents had a documented procedure.

94% of respondents stated that they had 'on-record' evidence of mineworker consultation in accordance with section 42(5) of the Regulation (which requires that the site senior executive must consult with a cross-section of workers at the mine in developing the fitness provisions for fatigue). 96% of respondents stated that they had developed the provisions in accordance with section 42(6) of the Regulation (which requires that the site senior executive provide a copy of the draft fitness provisions to the coal mine workers with whom (s)he consulted).

78% of respondents stated that they had 'on-record' evidence that the criteria for the assessment had been made in agreement with a majority of the workforce in accordance with section 42(6A) of the Regulation.

Fatigue related control strategies in use include:

- Hours of work and shift arrangements - There is a multitude of shift rosters, standard and maximum work hours per shift, maximum hours per shift cycle being used.
- Combating on-shift fatigue - 69% of respondents stated that they had proactive systems in place to control on-shift fatigue, including job rotation, extended rest and encouraging 'power napping'.
- 98% of respondents stated that they had an employee assistance program in place for persons who had persistent problems with 'on-shift' fatigue.
- Long distance commuting. 'Long distance commuting' is generally considered to be greater than one hour or 100 km of travel. 64% of respondents stated that they had a rule/policy in place regarding long distance commuting prior to and post roster cycle. For drive in/drive out operations, in particular, the accommodation arrangements are such that persons can have adequate rest pre and post roster cycle.
- Other fatigue monitoring and control factors - 56% of respondents stated that they utilized such devices as fatigue likelihood calculators and retina scanning technology and/or incorporated circadian rhythm and sleep apnoea research into their fatigue management planning.
- 37% of respondents stated that they had in place a policy and/or a system of disclosure of mineworkers involved in working off-site 'second jobs'.

Other physical and psychological impairment (PPI)

With regard to the SHMS for other physical and psychological impairment, 49% of respondents rated themselves "4 or higher". The lower implementation (compared to the other three fitness provisions) of these systems is attributed to the fact that they are more in the medical fitness realm covered by coal mineworker medicals and nominated medical advisor activities.

88% of respondents stated that they have, or are in the final stages of implementing a PPI protocol. 92% of those that are implemented treat physical impairment and psychological impairment as separate issues. They have a defined and agreed criteria in place for identifying, classifying and allocating normal or 'restricted' duties, which would normally involve medical specialists and risk assessment processes.

Some sites had difficulty defining what fitness provisions for physical and psychological impairment actually means and thus how to put in place effective management programs. Physical and psychological impairment tend to be classified together whereas they should be treated as two very separate classes of medical condition.

Education programs

Mine worker fitness provisions education was stated to be via pre-start induction and refresher training and toolbox meetings. Work force education was conducted in the following fitness provisions:

- Alcohol (96%)
- Drugs (89%) and
- Personal fatigue (96%).

The lower response for drug education is attributed to the fact that a number of the respondents are currently reviewing their site procedure with consideration of Recognised Standard 07 and a change to oral fluid testing.

Employee assistance programs

All respondents stated that their organisations (meaning mine operator employees) have an employee assistance program in place for their employees. In the main, this was stated as an appropriately qualified person or organisation, independent of the employer, providing assistance or counselling to employees (and in many cases, their families) for work and personal related problems. These may include fitness for work issues related to excessive consumption of alcohol, drug habits, fatigue and other physical and psychological impairments. A factor that was inconclusive in the compiled results was the number of mines where contractor and sub-contractor employers do not have employee assistance program for their employees. On the basis of those respondents who marked the "Not all" box in the alcohol and drug sections, this could be up to 20% of mine sites.

Training of supervisors to recognise FFW symptoms

Around two thirds of front line supervisors were stated as having received specific training in recognising symptoms of specific fitness issues. This included Alcohol – 70%, Drugs – 65%, Fatigue – 71% and PPI - 58%. The true positive response is probably somewhat lower than these figures due to the fact that some respondents misinterpreted this question that their supervisors had received training in the site's FFW systems 'full stop', not specific training in "recognising symptoms" of FFW issues.

Contractor fitness systems

91 % of respondents stated that their contractors operated under the site SHMS for FFW provisions, 5% operated under their own and 4% under both. In all cases, SSE approval was required and given for those working under their own system.

QME Recognised Standard 07

The original purpose of the Queensland Mines and Energy Recognised Standard 07 Criteria for the assessment of drugs in coal mines was to be used when an SSE could not gain agreement from a majority of mine workers on the criteria for assessment for the improper use of drugs. However it would appear that many sites have now adopted the criteria in Recognised Standard 07 for assessment fully or partially into their reviewed fitness provisions for drugs.

The Recognised Standard has recently been reviewed. The new standard puts greater emphasis on oral fluid testing as this testing is not as intrusive as urine testing and it provides information about recent drug use whereas urine testing provides information about longer term drug use. As recent drug use is more likely to relate to possible impairment oral fluid testing is considered more appropriate. However it should be noted that the Australian Standard states that oral fluid testing should not be used to determine impairment. The Recognised Standard therefore provides a testing mechanism rather than specific criteria for assessment. However as the mere presence of drugs above the cut-off or target concentrations denoted in the Australian Standard for oral fluid testing and for urine testing is considered a positive result, by default this provides criteria for assessment.

8

Mine Planning

Assuming that the reserve to be mined has been delineated with diamond-drill holes, the items listed in the following paragraphs need to be established with respect to mine planning for the mineralized material. Although it would be nice to have all of the following information before beginning every project or mine design and development, much of this information remains very speculative at the start of a project. One can work with only the information available.

Only the items preceded by an asterisk (*) are considered critical information needed prior to the initial development. There is a limit as to how much it is prudent to spend and how much time should be taken to determine the information while still at the stage of surface exploration drilling but having already established that the mineralization definitely is minable, General knowledge of similar rock types or structures in existing mining districts always is helpful. In developing the first mine in a new district there is far more risk of making costly errors than in the other mines that may follow.

GEOLOGIC AND MINERALOGIC INFORMATION

The geologic and mineralogic information needed includes the following:

1. The size (length, width, and thickness) of the areas to be mined within the overall area to be considered, including multiple areas, zones, or seams,
2. The dip or plunge of each mineralized zone, area, or seam, nothing the maximum depth to be mined,
3. The continuity or discontinuity within each of the mineralized zones.
4. Any swelling or narrowing of each mineralized zone.
5. The sharpness between the grades of mineralized zones within the material considered economically minable.
6. The sharpness between the ore and waste cutoff, including (a) whether this cutoff can be determined by observation or must be determined by assay or some special tool; (b) whether this cutoff also serves, as a natural parting resulting in little or no dilution, or whether the break between ore and waste must be induced entirely

by the mining method; and (c) whether the mineralized zone beyond (above or below) the existing cutoff represents submarginal economic value that may become economical at a later time.

7. The distribution of various valuable minerals making up each of the minable areas.
8. The distribution of the various deleterious minerals that may be harmful in processing the valuable mineral.
9. Whether the identified valuable minerals are interlocked with other fine-grained mineral or waste material.
10. The presence of alteration zones in both the mineralized and the waste zones.

Structural Information (Physical and Chemical)

The needed structural information includes the following:

1. The depth of cover.
2. A detailed description of the cover including: the type of cover; * the structural features in relation to the mineralized zone; * the structural features in relation to the proposed mine development; and *the presence of and information about water, gas, or oil that may be encountered.
3. The structure of the host rock (back, floor, hanging wall, footwall, etc.), including: * the type of rock; *the approximate strength or range of strengths; *any noted weakening structures; *any noted zones of inherent high stress; noted zones of alteration; the porosity and permeability; *the presence of any swelling-clay or shale interbedding; the rock quality designation (RQD) throughout the various zones in and around all of the mineralized area to be mined out; the temperature of the zones proposed for mining; and the acid generating nature of the host rock.
4. The structure of the mineralized material, including all of the factors in item 3 plus: *the tendency of the mineral to change character after being broken (*e.g.*, oxidizing, degenerating to all fines, recompacting into a solid mass, becoming fluid, etc.); *the siliceous content of the ore; the fibrous content of the ore; and the acid-generating nature of the ore.

ECONOMIC INFORMATION

The needed economic information includes:

1. The tons of the mineral reserve at various grades in all of the mining zones, areas, or seams. This is listed according to proven, probable, and inferred ores.
2. *The details on the land ownership and/or leaseholdings, including royalties to be paid or collected.

3. *The availability of water and its ownership on or near the property.
4. *The details of the surface ownership and surface structures that might be affected by subsidence of the surface.
5. *The location of the mining area in relation to: *any existing roads, railroads, or navigable rivers; *power; the community infrastructure; and *available commercial supplies.
6. The local, regional, and national political situations that have been observed with regard to the deposit.

PLANNING RELATED TO MINING SIZE

SIZING THE PRODUCTION OF A MINE

There is a considerable amount of available literature on the selection of a production rate to yield the greatest value to the owners. Basic to all modern mine evaluations and design concepts is the desire to optimize the net present value or to operate the property in such a way that the maximum internal rate of return is generated from the discounted cash flows.

Anyone involved in the planning of a new operation must be thoroughly familiar with these concepts. Equally important is the fact that any entrepreneur planning a mining operation solely from the financial aspects of optimization and not familiar with today's problems of maintaining high Ievels of concentrated production at a Iow operating cost per ton over a prolonged period is likely to experience unexpected disappointments in some years with Iow (or no) returns.

Other aspects of the problem of optimizing mine production relate to the potential effect of net present value. Viewed from the purely financial side, producing the product from the mineral deposit at the maximum rate yields the greatest return. This is due to the fixed cost involved in mining, as well as the present-value concepts of any investment. Still, there are "... practical Iimitations to the maximum intensity of production, arising out of many other considerations to which weight must be given".

There can be many factors Iimiting mine size, some of which are Iisted here:

1. Market conditions and the price of the product.
2. The grade of the mineral and the corresponding reserve tonnage.
3. The effect of the time required before the property can start producing.
4. The attitude and policies of the Iocal and national government and the degree of stability of existing policies, taxes, and Iaws that affect mining.
5. The availability of a source of energy and its cost.
6. The availability of usable water and a method of bringing in supplies and taking out production.
7. The physical properties of the rock and minerals to be developed and mined.

8. The amount of development required to achieve the desired production related to the shape of the mineral reserve.
9. The size and availability of the work force that must be obtained, trained, and maintained.

Market Condition and Price

Normally, the individual working on mine design is not the person doing the market analysis. To digress very far from the subject of mine planning to the subject of market effects on the size of mine production is not justified in this section-the problems do not relate to any particular type of mining but rather to the product produced. However, there are a couple of points concerning the elastic nature of markets that do relate to mine design. One of the advantages of the room-and-pillar mining-method is that it is extremely flexible in operation, and the mine can react to market needs faster than with other mining systems.

Most room-and-pillar mines have multiple faces for each unit or section to work. Where there is considerable irregular Iateral extent to the orebody, this could involve as many as 15 to 20 faces, or as few as one or two very Iarge faces in each section. For short periods of time in each stope unit, it usually is possible to work only the best (or poorer) faces, depending upon the market. This usually has a drastic effect on the grade within a few days. Similarly, spare equipment can be put into reserve stopes to increase production if the remaining materials flow can take the added capacity. However, if these practices are carried on too long or too often, mine development also has to be accelerated. If maintained, old stopes can be reactivated quickly to mine lower-grade minerals that become minable because of economic cycles. Even in room-and-pillar mining, changes in the rate of mining (momentum) cannot be assumed to be free. It often takes several months with an increased labor force to regain a production level that seemed easy to maintain before a cutback. If spare equipment is used, maintenance probably will convert to a "breakdown" overtime schedule compared to the previous preventive maintenance schedule on shift, at least until permanent additional equipment can be obtained. Nevertheless, the necessary changes can be made.

Another item often overlooked by individuals not accustomed to planning noncoal room-and-pillar mining is the situation in which mineral values are gradational. There is the option of mining through the better areas of the mineral reserve and maintaining a grade of ore that satisfies the economic objectives at that time. At a later time, when the mining economics may have changed, the lower-grade areas left as remnant ore reserves can be mined while slabbing or removing high-grade pillars. Thus, even in the latter years of the mining operation, some of the "sweetener" is Ieft to blend with the lower grade. Although not unique to room-and-pillar mining, this technique certainly is easier to accomplish in a room-and-pillar operation than in other more complex mining systems.

Grade and Reserve Tonnage Affecting Mine Size

No magical or statistical formulae can be applied to the parameters of grade (or value) of the mineral plus the tonnage available for mining to determine the ideal rate of mining, the proper mine life, or the best ultimate recovery of mineral values. For many reasons, such a determination is a complex problem.

First, the mine Iife and the rate at which the product is produced must yield the desired rate of return on the investment. So, the mineral reserve must satisfy at least these conditions. However, unless mining a totally homogeneous mass, it may make a considerable economic difference as to which portion is mined first and which is mined last. Furthermore, no ore reserve has an absolute fixed grade-to-tonnage relationship; trade-offs always must be considered. In most mineral deposits, lowering the mining grade (or quality) cutoff means there will be more tons available to mine. Even in bedded deposits such as coal, the ability or willingness to mine a lower seam height may mean that more tons eventually can be produced from the reserve. In such cases, the cost per unit of value of the product generally increases.

In considering the economic model of a new property, all of the variables of grade and tonnage, with the related mining costs, must be tried at various levels of mine production that, in the engineer's judgment, are reasonable for that particular mineral reserve. At this point in the analysis, the external restraints of production are ignored to develop an array of data that illustrate the return from various rates of production at various grades corresponding to particular tonnages of the reserve. At a later time, probability factors can be applied as the model is expanded to include other items.

TIMING AFFECTING MINE PRODUCTION

For any given ore reserve, the development required before production startup generally is related to the size of the production. For example, a very large production may require a larger shaft or multiple hoisting shafts, more and larger development drifts, simultaneously opening more minable reserves, and a greater lead time for planning and engineering all aspects of the mine and plant. In combination, all of these factors could amount to a considerable difference in the development time of a property. In turn, this would have two indirect economic effects:

1. The capital would be invested over a longer period of time before a positive cash flow is achieved.
2. The inflation rate-to-time relationship can pressure the costs upward by as much as 10% to 20% per year, thereby eliminating the benefits of the "economy of scale" of the large-size projects.

To aid the engineer in making rough approximations of the time parameters related to the size and depth of the mine shafts, Table 3.1 lists the average times for excavating, lining, and equipping vertical concrete circular shafts and slopes with a 0.26-rad (15°) slope for mines varying in size from 4535 to 18140 t/d (5000 to 20,000 stpd). The lateral development on each level of a room-and-pillar mine opens up new working places, and the mine-development rate can be accelerated each time a turnoff is passed, provided that there is enough equipment and hoisting capacity available. That is in contrast to mines that have a very limited number of development faces per level but more levels.

The timing of a cost often is more important than the amount of the cost. Timing is an item that must be studied in a sensitivity analysis of the financial model for the mine being planned. In this respect, any development that can be postponed until after a positive cash flow is achieved without increasing other mine costs certainly should be postponed. The flexibility of room-and-pillar mining often allows that to be done.

GOVERNMENT ATTITUDES, POLICIES, AND TAXES AFFECTING MINE SIZE

Government attitudes, policies, and taxes generally affect all mineral extraction systems and should be considered as they relate to the mining method and the mine size. Assume that a mine is being developed in a foreign country and that the political scene currently is stable but is impossible to predict beyond five to eight years. In such a case, it would be desirable to keep the maximum amount of development within the mineral zones, avoiding development in waste rock as much as possible. That would maximize the return in the short period of political stability. Such a practice usually is possible in a room-and-pillar mine, but it may result in a less efficient mine during later years. Also, it might be desirable to use a method that mines the better ore at an accelerated rate to get an early payback on the investment. if the investment remains secure at a later date, the lower-grade margins of the reserve then might be exploited. As a result of the flexibility of room-and-pillar mining, such a practice does not risk the recovery of the mineral remaining in the mine; block caving or longwall mining might risk such recovery. Similar situations might arise as a result of a country's tax or royalty policies, sometimes established so as to favor mine development and provide good benefits during the early years of production; in latter years the policies change. That would have the same effect as the preceding case; again, the flexibility of the mining rate and system must be considered.

PLANNING RELATED TO PHYSICAL PROPERTIES

The physical nature of the extracted mass and the mass left behind are very important in planning many of the characteristics of the operating mine.

Four aspects of room-and-pillar mining are particularly sensitive to rock properties:

1. The competency of the rock mass determines the unsupported open dimensions unless specified by government regulations. It also determines whether additional support is needed.
2. When small openings are required, they have a great effect on the productivity, especially in harder materials for which drill-and-blast cycles must be used.
3. The hardness, toughness, and abrasiveness of the material determines the type and class of equipment that can extract the material efficiently.
4. If the mineral contains or has entrapped toxic or explosive gases, the mining operation will be controlled by special provisions in the government regulations.

Preplanning from Geologic Data

Using geologic and rock-property information obtained during preliminary investigations, isopach maps should be constructed to show the horizons to be mined and those that are to be left as the roof and floor. Such maps show variances in the seam or vein thickness and identify geologic structures such as channels, washouts (wants), and deltas. Where differential compaction is indicated, associated fractures in areas of transition should be examined. Areas where structural changes occur might be the most favored mineral traps, but they usually are areas of potentially weakened structures. Where possible, locating major haulage drifts or main entries in such areas should be avoided; if intersections are planned in these areas, they should be reinforced, as soon as they are opened, to an extent greater than that ordinarily necessary elsewhere in the opening.

Extra reinforcing (or decreasing the extraction ratio) also may be necessary in a metal mine where the ore becomes much higher in grade than is normal, therefore having less strength. Where the pillars already have been formed prior to discovering the structural weakness, it probably will be necessary to reinforce the pillars with fully grouted reinforcing bars or friction stabilizers. It is advisable to map all joint and fracture information obtained from diamond-drill holes and from mine development, attempting to correlate structural features with any roof falls that might occur.

Room and Support Planning Related to Productivity

In relation to the stress imposed upon the material, the rock-mass strength of the material left behind in the mine has a definite relationship to the opening size and, therefore, to the mine productivity. This relationship is felt more in noncoal, nongassy room-and-pillar mines than in mines classified as gassy. The gassy mines arc subject to regulations of the Coal Mine Health and Safety

Act that designate the maximum span of the open room; other types of mines are permitted to use good engineering practices to establish optimum but safe room widths. The more competent the rock mass, the larger the openings can be for any given thickness and depth of a mineral reserve. Generally, a larger room allows greater maximum production per day from that face and greater productivity from the workforce. However, because other factors affect the overall mine productivity, this generalization is not reliable. Some productivity relationships, indicating a degree of correlation between the face area and the corresponding percentage of extraction and resulting productivity.

The last one, were breaking and loading rock of a similar nature. There are reasons why the limestone mines can open rooms that usually are much larger than the metal mines, but that is of little importance here. The important indication is that the mine engineer should seek to open rooms as large as possible to achieve maximum productivity without causing problems with grade control, safety, or rock stability.

Hardness, Toughness, and Abrasiveness of Extracted Material

The hardness, toughness, and abrasiveness of the material determine whether the material is extracted by some form of cutting action (raking, plowing, or scraping), by drilling and blasting, or by a combination of both methods. Technological advances in hard-metal cutting surfaces, steel strengths, and available thrust forces allow increasingly harder materials to be extracted by continuous mining machines. The economics of continuous cutting or fracturing as compared to drilling and blasting gradually are being changed for some of the semihard or tough materials. However, for continuous mining to be competitive with modern high-speed drills and relatively inexpensive explosives, it appears that the rock strengths must be less than 68,900 to 103,400 kPa (10,000 to 15,000 psi).

At times, reasons other than simple economics favor the use of one mining system over another. A tunneling machine may be advantageous in protecting the remaining rock where blasting might be prohibited. Drilling and blasting may be necessary to avoid pulverizing a shale parting between two coal seams. A continuous boring tool may be desirable for totally extracting an orebody without personnel having to enter the stoping area.

Material Containing Gas

Whether the mine is classified as gassy drastically affects the relationship of noncoal hard-rock mine design, efficiency, productivity, and overall profitability. A gassy mine is more regulated by the Federal Mine Health and Safety Act. It would be difficult, if not totally uneconomical, to operate any hard-rock mine (regardless of the material being mined) that must drive multiple entries where only single entries are needed and that has to follow all other current specia interpretations of a mine with a gassy classification as they apply to coal.

Rock Strata Conditions-Effects of Unknown Factors

Any unexpected and abnormal condition creating a stability problem in a stope or pillar or causing excessive inflow of water or gas undoubtedly causes the mine to produce less than its design capacity, at least temporarily. Although these problems may be solved, encountering them unexpectedly may delay reaching full capacity until the problems are corrected.

If the ground is much worse than expected, the back and pillars may have to be reinforced more than planned, requiring time and resulting in a temporary loss of production. The size of the rooms may have to be reduced, again decreasing the productivity of that unit. Maintaining planned production may require the addition of more units and an acceleration of the development effort.

Because the development headings of room-and-pillar mines usually are well advanced beyond the operating faces, it is the development drifts or entries that normally receive most of the water or gas inflows. Sufficient pumps must be provided to handle the water, or sufficient ventilation must be provided to dilute or remove the gas. If the mine has reached the limit or capacity of either pumping or ventilation systems, the development advance then must be slowed to accommodate the problem. To avoid problems of this type, it is important to obtain as much information as possible before attempting to develop a mine in an unfamiliar area.

PLANNING THE ROOM-AND-PILLAR MINE

When describing room-and-pillar mining, there is some confusion as to the meaning of "production development." That confusion results from different interpretations in hard-rock mines and mines for softer materials such as coal, trona, salt, and potash. A development in a hard-rock mine is any initial opening driven to provide an access way anywhere in the mine. It may not necessarily be in the ore. Such access ways usually are driven as single openings and provide space for roads, ventilation, power, air and water lines, drainage, storage areas, shops, dump stations, offices, etc. They also expose the ore so that production mining can begin by opening rooms around pillars beginning at the point of exposure.

As used in these mines, the term "development" has nothing to do with whether the opening also is starting the rooms and pillars. It does not provide any indication of whether the pillars will be recovered while advancing or retreating, and it does not indicate whether the mine is advancing a production face in virgin ground that may or may not have other slices taken either above or below it. The term implies only that there is an opening for some mine service or where production might begin. In contrast, basic coal-mining techniques use the term "development" to refer to that phase of the mining cycle during which multiple entries are advanced through a virgin seam, forming pillars by breaking crosscuts or breakthroughs between the entries. When the development has reached the extent of the advance, the coal-extraction process removes all or part of the pillar on the retreat. The only area of obvious similarity between the two meanings is where the development opens more minable mineral.

Some confusion also exists between the two mining applications when describing the initial entrances into the room-and-pillar mine. With coal-mining techniques, "drift entries" are driven nearly level into the side of a hill, and "slopes" are driven downgrade at an angle too low for the material to move by gravity. In hard-rock mines, the initial entrance driven nearly level into the side of a hill is known as an "adit"; if driven downgrade, it is known as a "decline" or "ramp." If an opening is driven upgrade at an angle too low for the rock to move by gravity, it is known as an "incline" or "ramp." In any mine, an opening driven upward at an angle steep enough for the rock to move by gravity is known as a "raise." A vertical opening sunk into the mine from the surface is known as a "shaft."

In metal mines, underground openings parallel to the general trend of the main structure are known as "drifts," but when driven at a right angle or a high angle to these structures or drifts, the openings are known as "crosscuts." In coal mines, the breakthroughs between entries forming the pillars sometimes are known as "crosscuts." The two sets of terminology developed as a result of little past association between proponents of the two broad categories of room-and-pillar mining. However, this has changed, and many companies now operate both types of mines, and many of the equipment manufacturers now sell and service both groups. Furthermore, the same government research and enforcement agencies work with both types of mines.

The use of the same government agencies for both types of mines is one clue to the underlying problem of trying to administer a common interpretation of laws that have been developed by people understanding only one mining system and not really understanding the differences or the purpose for which the mine opening might be driven in the other system. For example, a single development in a limestone or metal mine may serve any of a number of purposes, none of which is the same as or has any resemblance to the same conditions that require service in a coal mine. Still, some people attempt to apply the same rationale to these extreme conditions, forcing an unjustified hardship onto all of those involved.

It is extremely difficult to describe each and every type of mine development because of the terminology confusion and because openings can be driven in any direction, in all sizes, and by many different methods. An attempt to show that there are three basic reasons for which all developments are driven. For each development, decisions must be made pertaining to each of the five physical variables. In the final analysis, each development must serve the purpose for which it was intended and usually named.

DEVELOPMENT DRIVEN FOR MINE-PLANNING INFORMATION

During the very early stages of evaluation and planning, a great deal of information is needed to make the correct decisions concerning the property. Therefore, the initial opening might be driven to test a particular aspect of the mineral reserve that is considered critical for successful exploitation. Most of the technical information listed in the preceding paragraphs can be obtained from surface diamond-drill holes, particularly where the mineralized zone is fairly shallow, has continuity, and is flat.

Unfortunately, it is common that the true economic potential of many mineral reserves cannot be determined until after an exploration development is driven. From the development opening, underground prospecting can further delineate the materials present, a bulk sample can be taken for metallurgical testing, and/or the true mining conditions can be revealed. This procedure involves considerable financial exposure, and, unfortunately, the delineated reserve may turn out not to be economically suitable.

Nevertheless, there may be Iittle choice about driving a development opening if the potential of the reserve is to be proven and the risk to Iarger downstream capital expenditures is to be minimized. In contrast to deep-shaft development for a test mine, it often may be possible and more economical to drive the development opening as an adit or slope.

Regardless of the reason for driving a preliminary development, three basic principles should be observed:

1. The development should be planned to obtain as much of the critical information as possible without driving the opening in a Iocation that will interfere with Iater production development. In room-and-pillar mining, it usually is possible to Iocate at Ieast a portion of the exploration development where it can be utilized in the developed mine, although it Iatter may be necessary to enlarge the openings to accommodate the production equipment.
2. The preliminary development should be held to the minimum needed to obtain the necessary information, and the development should be completed as quickly as possible. Most exploration drifts for metal mines are only large enough to accommodate small loading and hauling units [from 2.4 x 3.0 m to 3.0 x 4.3 m (8 x 10 ft to 10 x 14 ft)] .If geologic information is needed, it often is possible to Iocate the developments in close proximity to several of the questionable structures, without actually drifting to each one. With modern high-speed diamond drills having a range of 213 to 305 m (700 to 1000 ft), these structures usually can be cored, assayed, and mapped at minimum cost.

3. Once the preliminary development has been done to satisfy a specific need, the development should be used to maximize all other physical and structural information to a maximum extent that will not delay the project.

There are no rules governing the size of the shaft or entry for exploration development. The size depends upon many factors, including the probability of mine development to the production stage. If the probability favors follow-up with full mine development immediately after exploration development, and if the shaft depth is not excessive [*e.g.*, 244 to 366 m (800 to 1200 ft], it would be logical to sink a shaft sized for Iatter use in production development. This is particularly true of operations normally using small shafts for production.

In contrast, some companies choose smaller drilled shafts to speed shaft sinking and minimize early expenses. Such shafts usually are completed considerably faster than conventional shafts but not as fast as raise boring (where an opening already exists, the hole is reamed upward, and the cuttings fall back into the mine). Because this section is directed toward initial mine openings, cuttings removal must be upward through the hole and out the collar; various methods use air or mud with either direct or reverse circulation. Of the 83 shafts surveyed, amounting to 16,587 linear m (54,420 Iinear ft) of boring, 58% were blind drilled. Although bored shafts stand very well without support, 58% of the surveyed shafts contained either a steel or a concrete Iining.

Recently, shotcreating bored shafts has become acceptable and probably results in substantial cost savings. Of the 83 shafts, 73% were drilled 1.8-m (6-ft) diam or smaller. Of all the blind-drilled shafts, 55% were bored 305 m (1000 ft or less), and 39% were over 305 m (1000 ft) but Iess than 550 m (1800 ft) .Only one shaft was over 610 m (2000 ft) .In the survey of conventional shafts developed between 1963 and 1973, 41 of the mines were room-and-pillar mines, and 88% of these were less than 550 m (1800 ft) deep. These data suggest that many of the mines were within reach of blind-bored shafts if other conditions such as rock hardness and abrasiveness were amenable and a small diameter shaft could have been utilized for development.

Developing the Production Access Way and System

The basic considerations in planning and developing the production access way and system are related directly to the design parameters of the development openings. These are the function in the production flow, the inclination of the opening, the size of the opening, and the arrangement and support of the opening.

The Production/Development Function. Each element of the production flow process must be considered, and each must be planned to be compatible with the elements of flow on both sides of it. In a room-and-pillar mine, a typical sequence of these elements in the development openings for material flow from the production face might be a preliminary load-haul-dump drift, an orepass, the main haulage drift, the main ore pocket, an orepass to the crusher-skip loading station, and a shaft for hoisting to the surface. Although other configurations and production flow sequences are used, the preceding sequence is typical of noncoal room-and-pillar mining in the United States. In the following paragraphs detailing inclination, size, and arrangement, the production developments from the surface are considered first, followed by descriptions of the internal room-and-pillar mine developments. Inclination of Developments from the Surface.

As described earlier, the inclination determines the name of the development. The inclination is determined by a combination of the relative elevations of the surface and the mineralized zone, and the economics of driving the development and operating the production system through the development. Drift Entry (Adit). An adit, or drift entry, is the most economical approach when the minable material is above the floor elevation of a nearby valley. Literally thousands of Appalachian room-and-pillar coal mines have been opened this way to mine coal seams exposed on the hillsides; these are known as "punch mines."

When coal was mined in the west to power steam locomotives, this same type of mine was known as a "wagon mine." Most of the early western mines used this type of development, as did most of the limestone and dolomite mines throughout the Missouri-Mississippi river basin. The exposed ore could be opened through horizontal adits into the sides of the mountain bluffs. When the minable material goes to a depth lower than the surrounding terrain, either a slope (decline) or a shaft must be developed. Slopes (Declines) vs. Shafts. There is no general rule governing whether a shaft or a slope is the proper type of entry to be developed; each has advantages and disadvantages. Slopes usually are limited to relatively shallow mines because, for a given vertical depth, they require approximately four times the linear distance required for a shaft.

Thus, to reach a depth of only 460 m (1500 ft) a normal slope would be approximately 1840 m (6000 ft) long. Although the cost per unit distance for a decline may be somewhat lower for a slope than for a vertical shaft, the overall cost usually is higher an the development time usually is longer as a result of the greater length required. Most new shafts being sunk today are concrete lined, offering excellent ground stability. Most slopes have a rectangular shape, with the widest span for the floor and roof. In incompetent ground, slopes can be very difficult to develop and very difficult to maintain over the life of the property. In ground that must be frozen prior to excavation, a slope requires freezing a much larger volume of material than does a shaft, and it undoubtedly would cost more for a slope than for a shaft. However, slopes do offer some advantages:

1. Where conveyor-belt haulage brings the material to a central point, it is logical to continue haulage to the surface by means of a belt conveyor up a slope. Where slopes and conveyor systems are used to move the material to the surface, the mine generally is geared for a consistent high rate of production.
2. Slopes provide easier access for mobile equipment entering or leaving the mine; in many cases, the equipment can be transported or driven into the mine intact. With a conventional shaft, large trackless equipment may have to be disassembled completely and then reassembled underground; that is particularly difficult when the unit is a large truck, and the bed must be split and rewelded. A similar advantage is found when using rail haulage in a slope for transportation of supplies-the material can remain in its original container for transport from the surface.
3. Where minor ground movements are a problem, a production slope offers considerably less difficulty in maintenance than does a vertical production shaft.

Despite the advantages of slopes, the disadvantages of cost and development time appear to have the greater influence on noncoal mines. Although most noncoal mines using room-and pillar mining are reasonably shallow, vertical shafts continue to be used for their development. In a l0-year period around 1970, only four noncoal mines had been developed by declines, and three of those were room-and-pillar mines (Dravo Corp., 1974). The depths varied from 105 to 590 m (345 to 1940 ft) with grades varying from 0. 17 to 0.35 rad (10° to 200) and averaging 0.26 rad (14.7°). For that time period, this represents only 9% of the total linear footage of noncoal mine development from the surface.

Four exceptions did not fit into the survey mentioned above: the White Pine copper mine of Copper Range Co., the Maysville mine of Dravo Corp., the Lost Creek mine of New Jersey Zinc Co.., and the Gordonsville mine of New Jersey Zinc Co. Of these, the first three use conveyor belts to transport the material from the operating areas to the surface. The economics of using declines become obvious only where conveyor belts are used throughout the operation or where the mine is very shallow and rubber-tired vehicles can be driven from the working area to the surface. Many small mines, close to the surface, also are opened up with declines; the mines found in the Illinois-Wisconsin zinc district are typical of this type of entry.

Size of the Opening. The size of the opening should be tailored to accommodate the specified purpose. For production shafts, the shaft cross section is influenced (but not dictated) by the skip capacity, and the skip capacity is determined by the variables of mineral density, hoist capacity, hoisting speed, skip loading time, skip dumping time, and allowable hoisting time.

Arrangement of Facilities in the Production Opening. The production shaft or slope must accommodate the conveyances for production material and may, at times, also contain some of the services such as an emergency man cage and/or ladderway. Whether such services are located in the production opening is dependent upon the number of other openings from the mine to the surface. Usually, the service facilities are contained inside pipes that may be embedded in the concrete shaft lining.

Most new conventional shafts are lined with 0.3 to 0.6 m (1 to 2 ft) of concrete. However, there are cases in which only part of the shaft is lined. Timbering for support of shafts generally is limited to deepening older rectangular shafts or to small mine openings not using mechanized shaft sinking. Hoisting in a shaft is accomplished either with two containers (skips) that balance each other or with one large skip that is balanced by a counterweight.

Most slopes (declines) utilize roof bolting and/or shotcrete with a concrete portal. The three arrangements for production slopes are: (1) one open compartment, (2) two side-by-side compartments, or (3) two over/under compartments. Drifts, Entries, and Crosscuts for Production. Although the following information primarily addresses noncoal room-and-pillar mines, many of the concepts apply to basic mine development planning for other types of mines.

The configuration of a room-and-pillar mine generally follows the shape of the mineralized orebody to be mined. Therefore, the basic shape of the mine should start by determining the path of the main haulageways (entries), branching from these to reach the lateral extent of the reserve. Wherever possible, a few basic principles should be followed in planning the production developments. The central gathering point for hoisting should be as near to the center of the minable mass as practical. That keeps the haulage, personnel, and material movement distances to a minimum. Although this may mean that a large shaft or slope pillar is left in the center of the mineral reserve of a flat-lying deposit, it does make the mining operation more efficient. At the end of the mining project, most of the pillars probably can be extracted.

Normally, room-and-pillar mines without ground-stability problems should be developed as symmetrically as possible to allow the maximum number of working areas to be opened with a minimum of development time. In trackless mines, it is advisable to cut the main haulage opening as much as 50% larger than the minimum size needed to operate the existing equipment. Large main haulageways improve productivity, ventilation, and road conditions. They also allow productivity to be maintained as the haulage distance to the shaft increases by permitting efficient operation of larger equipment. Haulage roads and railroads should be kept as straight as possible, with steep grades other than conveyor slopes being kept as short as possible. The grade limits depend upon the type of haulage equipment being used.

For main haulageways, railroad grades should be less than 3% and it is advisable to keep the grade less than 8% for rubber-tired diesel equipment. However, for very short distances where the equipment is allowed to gain mass momentum, 15% to 18% grades can be negotiated quite efficiently. If a 20% decline is to be driven for a conveyor system or service slope, rubber-tired diesel-powered loaders are the best equipment for loading and hauling the material from the development.

The radius of a curve that can be negotiated by trackless equipment is somewhat dependent upon the size of the equipment in relation to the size of the drift and the turning radius of the equipment. Normally, a radius of 15 m (50 ft) is handled easily by any trackless mining equipment; unless it is articulated, a road grader usually requires the greatest turning radius. For rail-mounted haulage, a radius of 24 m (80 ft) is adequate for stope development, but a limited-service main line requires a minimum radius of 38 m (125 ft).

However, a radius of 60 to 150 m (200 to 500 ft) would be appropriate for main-line haulage, depending upon the speed of the train. Preceding sections have outlined the need for operating flexibility in relation to the size, shape, and thickness of the mineral reserve. Other aspects of the mine also require flexibility in the plan. For example, it often is advantageous to incorporate a lower level of uniform grade below the main ore horizon to create a ventilation circuit, providing mine drainage that removes water from the operating headings and roadways to make ore movement more efficient. It is important for the mine planner to remember that the main haulage drifts will be used for all of the ore produced in the mine.

Therefore, these drifts should be planned carefully for optimum production. If the ore can be moved quickly to an efficient main transportation system, it will not matter nearly as much if the stope developments are small or crooked. Raises and Ramps as Production Developments. The use of raises and ramps as production arteries also must be planned. If the ore is to be transferred vertically to the main transportation artery, two different approaches may be used: either the ore is dumped into an orepass where it moves by gravity and can be reloaded at a lower level, or the ore is kept in the same conveyance for transport down or up a decline or ramp to the desired location.

In modern rail or conveyor-haulage mines, material from the stopes usually is dropped through an orepass to reach a lower gathering point. To move ore up a grade that is too steep for self-propelled track or trackless haulage, the ore either can be hoisted up an incline in individual rail cars, or it can be conveyed up an incline for subsequent transfer into rail cars or onto another conveyor. In modern mines using trackless haulage throughout, the practices are mixed-some operators use steep ramps, whereas others use orepasses and a reloading system. The advantages of not having to reload the material include no time lost for dumping and reloading, no extra mine development for an orepass, no extra cost for installing a reloading system, elimination of the high-risk chute operator's position for reloading, and maintenance of less equipment.

Factors favoring the use of an orepass include no time lost traveling an inefficient curving ramp, no time lost negotiating steep grades, less danger to vehicle operators because haulage is on moderate grades, lower maintenance and operating costs for haulage vehicles operating on moderate grades, and the availability of automatic or semiautomatic chutes for loading trucks or belts. High-productivity feeders can load large trucks within 1 or 2 min, or front-end loaders can reload themselves quickly; a chute operator would not be required. The decision as to which method should be used must be based upon the tonnage eventually taken out of the area in relation to the criteria for efficient haulage.

Today's trend is to maintain very high productivity; if the haulage ramp is at all steep, most room-and-pillar mine operators favor orepasses and reloading if the tonnage justifies the installation. If ramps are used, it has been found that ramps with 3.14-rad (180°) turns and long straight sections are more efficient than continuous spirals. When planning raises for production, several critical factors must be considered: the size of material to pass through the raise and whether it will flow by gravity, the effects of passing ore on the size and stability of the opening, and the least-cost method of developing the raise in relation to its length. The same design problems become even more important in mining systems that rely heavily upon orepasses, chutes, or drawpoints.

Ore-Storage Pocket. Some surge capacity must be provided between the normal mine production gathered from various parts of the mine and the conveyance that transports the material to the surface. The need arises because the material usually has many production paths to reach the shaft, slope, or adit, but only one path out of the mine. Therefore, the single-path conveyance is vulnerable to downtime for maintenance and repair, but such downtime cannot be allowed to disturb the rest of the mining cycle. Similarly, the multi path production can operate intermittently (two shifts per day), while the material can flow from the mine continuously.

There is no general agreement on the optimum size of an ore pocket; both the nature of the operations and the management priorities differ. However, the two major considerations are: (1) whether there is storage at the discharge end of the hoisting operation before the ore proceeds to the next step in the flow, and (2) the size of this storage and whether interruptions before the next step are critical.

Another reason for using ore-storage pockets before hoisting is to keep various products or waste materials segregated for separate hoisting. The products may be different minerals, or they-may be the same mineral but owned by different people. The ownership might insist that the material be kept separate until after crushing, weighing, and sampling so that royalties on the value can be paid correctly. Such practices are quite common in mines where government ownership of the mineral is involved.

PLANNING THE SERVICE DEVELOPMENTS

Any mine opening is a service development if it is not developed specifically to gain information about the mineral reserve and mining conditions, to obtain a bulk sample, or to handle the production system. Service developments are subject to the same basic considerations as are the production, openings, *i.e.,* their functions, inclinations, sizes, and arrangements.

The name of the development is a combination of the function and the inclination as are the names of production developments. Thus, service developments include ventilation raises, personnel and materials service shafts, sump cleanout drifts, emergency escape shafts, etc. A service development may serve a combination of functions or change its function over the life of the property. Basically, the design rationale for service developments follows that of production developments, modified by special considerations associated with the service function.

Ventilation Openings. The current trend in ventilation raise and shaft development is for the opening to be bored or upreamed, rather than drilled or blasted. Such holes can be completed by an outside contractor or by the mine crew that may own or lease a boring machine. Most of the bored shafts were either 1,5 o 1,8 m (5 to 6 ft) diam; these may be too small to meet today's requirements. The trend should be to enlarge the opening size to reduce ventilation power consumption and to move the needed amount of air at minimum cost. That is problem of optimization that has to be solved for each individual user. In horizontal developments or slopes, ventilation tubing or ducting is used in most metal mines, and brattice cloth or canvas is used in coal mines.

Shop and Storehouse Developments. The efficiency of a modem mining system depends heavily upon the productivity and the availability of the equipment used to extract the material. Compared to other mining methods, little gravity movement of the ore is used in room-and-pillar mines. Therefore, most of the energy to move the material must be provided. Because the material may be very heavy and/or abrasive, and because the environment of an underground mine imposes adverse operating conditions, underground mining equipment requires a great deal of maintenance and repair. One of the most serious and most prevalent errors made in designing a mine is the failure to provide adequate space and equipment for necessary maintenance and repair work. The amount of service, if any, provided underground depends upon several factors:

1. The degree of difficulty in moving equipment into or out of the mine is a major consideration. If the property is a limestone mine with adit entrances and a good shop on the surface, it probably would not be advantageous to duplicate the facilities and personnel underground.
2. Underground shops should provide a safe and good working environment. In mines that are gassy or carry a gassy classification, building underground shop facilities may not be practical. However, there are the same advantages for gassy and nongassy mines; therefore, some room-and-pillar trona mines have very good, Iarge, and well-equipped underground shops even though the mines are classified as gassy.
3. If the active working area will be abandoned totally after a fairly short life moving to another such area nearby it would not pay to invest in an extensive underground shop area.

Most other room-and-pillar mines should develop good underground shop facilities. In the New Lead Belt of Missouri, where the mines are developed with shafts 152 to 396 m (500 to 1300 ft) deep, all of the mines rely upon large mining equipment and have well developed underground shop areas. The total area usually utilized as a shop area can range from 465 to 1394 m2 (5000 to 15,000 sq ft).

The important criteria for shop design are to provide:

1. One or two Iarge bridge cranes over the motor pits.
2. Easy access from several different directions for access to and around the cranes when they are in service for extended periods of time.
3. A motor pit and service area for scheduled Iubrications performed as part of the preventive maintenance program.
4. A separate area for welding operations.
5. A separate area and equipment for tire mounting and repair.
6. A separate area for recharging batteries.
7. Various work areas equipped with steel worktables or benches.
8. Close proximity to the main supply house.
9. An area for washing and steam cleaning the equipment.
10. An office for the shop foreman, records, manuals, catalogs, drawings, and a possible drafting table. The main pit areas should be visible from the office windows.

Sump Area and Pump Station: In mines below surface drainage, areas must be provided to store the water before it is pumped to the surface and discharged. There is no general rule of thumb to determine the capacity for water storage.

Adequate pumping capacity is the only permanent solution to removing water from a mine. However, fluctuations in the water inflow and/or the time periods when some or all of the pumps are inoperable must be handled with an adequate sump capacity. Sumps also are needed in wet trackless room-and-pillar mines; the rubber-tired vehicles traveling on the roadways create fine material that eventually collects in the water ditches if the roadways have drips or continuous streams of water on them.

Although this situation should be corrected to keep water off the roadways, it invariably occurs or reoccurs. A place must be provided for the fines to settle out of the water so that they do not damage the clear-water high-head pump impellers. That sometimes can be done effectively in a small catch basin that is cleaned out frequently. However, if the water flows are substantial, the main sump receives the bulk of the fine material. As a result, provisions must be made to clean the sump with conventional or special equipment, driven the water into another sump while one sump is being cleaned, provide a place to put the fines ("soup") removed from the sump, and devise a method of transporting the fines there.

One of the best methods of cleaning a sump is to use conventional front-end loaders to remove the material. However, a ramp must be provided down to the sump level floor. if deep-well impeller sections are suspended from a pump station down into the sump, considerable care must be taken to avoid bumping the impellers with the loader. Other systems using slurry pumps, scrapers, or diverting the material into skips have been tried, but there still is no good way to clean a sump.

Diverting the water flow into another sump while cleaning one sump also can be a problem unless at least twice the normally required pumping capacity is provided. Half of the pumps have to handle all of the inflowing water while one sump is being pumped down and cleaned. Mines that make large amounts of water also produce large amounts of fine material that fill the sumps much faster than anticipated. The sump of a new mine in the development stage cannot be cleaned too soon; if the mine is not producing much water at a particular stage of development, the sump still should be kept clean. The next shift may bring on more water than can be handled with half the pumps, and once that point is reached there will be no way to get to the muck in the bottom of the sump for cleaning.

Two methods often are used to divide the sumps for cleaning. One is to locate a concrete wall between the two areas and to provide a means of closing off one side and making the water flow to the other side. With this method, the wall should be anchored properly on the top and bottom with reinforcing steel into the rock. This provides a safety factor in preventing the wall from collapsing into the side being cleaned. The other system is to develop two physically separate sumps. Both sumps must be provided with pumps, a method of stopping the water inflow and diverting to the other sump, and a means of access for cleaning. Some mines locate the pump station below the sump chambers, using horizontal centrifugal pumps instead of deep-well turbine pumps. These mines avoid the installation of a vacuum priming system because the overflow feeds the main pump. Although this system provides easier access to the sumps and makes cleaning easier, the risk of losing the pump station by flooding is increased.

Other Service Developments. A variety of other rooms or drifts also must be developed. A lunch area or room is nearly essential around an underground shop. Office space should be provided for the privacy of the underground supervisor and the security of records and equipment. Ramps driven strictly for access are important. Normally, these can be driven steeper than haulage ramps because the equipment using them would not be loaded and time to negotiate them would not be as critical. However, the ramp must not be so steep as to prevent unloaded equipment from having good traction. Equipment manufacturers can advise the mine planner on this point.

A small area at the bottom of the shaft may have to be provided for the equipment to clean the shaft bottom. In some cases, equipment such as an overshot loader will have to remain on the ramp, so there must be sufficient maneuvering room between the shaft bottom and a bucket that lifts the loaded spill rock out of the bottom of the shaft. Ideally, but seldom, there is a drift intersecting the bottom of the shaft and a ramp leading back up to the bottom level of the mine. In such cases, a steel deflection door can be pushed into place for shaft cleaning in the access drift while hoisting is in progress. Shaft cleanout also may be accomplished by remote-controlled loading equipmention the shaft wall or on the steel below the skip-loading facility but above the shaft bottom.

PLANNING THE ORGANIZATION AND REQUIRED EQUIPMENT

The amount of equipment or personnel for the needs of all mines cannot be specified in general terms. However, personnel and equipment productivity ranges can serve as guidelines for planning.

WORK FORCE AND PRODUCTION DESIGN

It is necessary to consider several factors concerning the work force to operate the mine. Many questions needing investigation will be difficult to answer, but they have profound effects on the financial success of any mining project and eventually must be faced.

1. Is the supply of labor adequate to sustain the production level dictated by other economic factors? If not, can the needed labor be brought in, and at what cost?
2. What is the past history of labor relations in the area ? Are the workers accustomed to a five-day work schedule and, if so, how will they react to a staggered six- or seven-day schedule?
3. Are the local people trained in similar production operations, or must everyone be trained before production can achieve full capacity?

4. Can people with maintenance skills be attracted to the property, or will the maintenance crew have to be built up through an apprenticeship program?

Apprenticeship programs are very slow processes. Accordingly, some state laws restrict the number of people who can be trained each year in such programs. That one item could cause a mine designed and equipped for a very large daily production to fall far short of its goals.

It is important that the "people problems" be investigated at the same time that the property is being evaluated and designed. That provides adequate time for specialized training, minimizes unexpected costs, and prevents basing economic projections on policies that, if implemented, could destroy employee morale or community confidence. The productivity and profitability difference between an operation with good morale and good labor relations and an operation with poor morale and poor labor relations (with many work stoppages) can make the difference between profit or loss, expansion or curtailment, and go or stop. Of all the items involved in mine design, this one is the most neglected and can be the most disastrous.

One of the advantages of modem room-and-pillar mining systems is that every task can be mechanized to some degree, provided that it is economically sound to do so. That minimizes the operating labor force and makes manning the operation easier. The high-capacity equipment for modem room-and-pillar operations is reasonably simple to learn and to operate. Anyone having operated any heavy machinery in construction work, the military, or even on a farm has little trouble adapting to loading and hauling equipment in a room-and-pillar operation. if there is a nucleus to start from, there usually is help available from the equipment manufacturer in training maintenance people.

Equipping Room-and-Pillar Mines

Detailed information concerning mine equipment is contained in other sections of this handbook. The best way to increase productivity is to utilize equipment designed for that purpose, that is, to mechanize wherever a safe and reliable piece of equipments available to magnify the physical and mental efforts of an individual.

Equipment Versatility. The equipment selected for room-and-pillar mining should be as versatile as possible. For example, in one mining operation, the same high-performance rotary percussion drilling machines were used for drilling the bluff or brow headings and were mounted on standard drill jumbos for drilling holes for burn-cut rounds and slabbing rounds in the breast headings. Because the drills on these jumbos penetrate extremely fast, they also were used to drill the holes for roof bolts and, in some cases, to drill holes for reinforcing pillars. The same front-end Ioader was used to Ioad trucks in one stope and to perform as an LHD unit in another stope. By switching working platforms, the same forklift tractors served as explosive-charging vehicles and as utility service units for hanging air, water, and power lines. They also served as standard forklifts for handling mine supplies. That philosophy results in the following advantages:

1. There is less equipment to purchase and maintain.
2. Less training is required for operators and maintenance personnel. At the same time, all personnel have a better chance of becoming more efficient their jobs.
3. Having fewer types of machinery means having less inventory to obtain and maintain.

Some of the possible disadvantages of this philosophy include:

1. A more efficient machine may be available to do the job being done by the versatile machine. Therefore, if a great amount of that type of work is to be done, it may be advisable to use the specialized machine.
2. The mine may become too dependent upon a single source of equipment supply.

Application Flexibility. The selected equipment should be flexible in application. That is, the equipment should be able to accelerate and move rapidly, have good balance and control at high speeds, be very maneuverable, and have plenty of reserve power for severe applications. Both trucks and loaders should have ample power to climb all grades in the mine and to accelerate quickly to top speed on long straight hauls.

The minimum mobile equipment necessary to equip either a conventional or continuous coal-mining section. Where roofbolting is the "bottleneck" in the operation, as it usually is, the current trend is to use either two-boom roofbolters or two complete single-boom roofbolters. Where the seam height permits very high productivity from the loading machine or continuous miner, more than two shuttle cars may be required.

PLANNING PERSONNEL AND SUPPLY HANDLING

During the early stages of mine planning, consideration should be given to the methods and problems involved in transporting personnel and materials into the mine and to the working face. Because every operating room of a room-and-pillar mine has an access roadway that can be negotiated by mobile equipment, personnel and material transportation is considerably easier in these mines than in other types of mines that are on steeply dipping veins.

The following conclusions concerning excessive personnel and material transportation time came from a report;

1. Typically, equipment transport speeds are not up to the level of performance currently achievable through existing technology.
2. Nonmechanized means of travel such as walking and climbing are significantly slower than mechanized transport. Over long distances, such travel results in lost mining time. Walking and climbing have

been identified as major problem areas with respect to lost-time accidents in personnel movement. Underground environments typically make walking or climbing slow, tedious, and hazardous.

3. General mine conditions such as poor track and rough or crooked roadways adversely affect the operational efficiency of personnel transport systems.
4. In some cases, equipment design impedes efficient loading and unloading of personnel, thereby increasing the transportation time.

The same report lists the following recommendations as ways to improve personnel movement:

1. To optimize personnel transport, each mine should identify specific groups of personnel according to job functions, listing their ultimate destinations, intermediate transport requirements, etc.
2. The system component dependence should be handled by personnel scheduling techniques. The horizontal system should accept the personnel unit on a predetermined schedule.
3. An engineering study of all personal movement elements should be conducted to establish present and future requirements for a balanced transport system.
4. Crews should be transported to a point as close as possible to the actual work, and they should be picked up on a predetermined schedule at the same point for outbound transport.
5. Supervisory staff should use mechanized transport wherever possible.
6. Daily production reporting procedures should be utilized, rather than end-of-shift recording procedures.

An incentive or supervision system must be incorporated to assure that the time gained at the face is utilized productively. It cannot be assumed that putting a miner at his working place a given number of minutes early automatically will increase his daily output-the miner must be shown some personal benefit in actually utilizing those extra minutes.

MINE PLANNING AND DESIGN (MPD)

MPD approach

Simplistically, an underground mine may be considered to be an engineering structure that is made up of a two components:

- Mine voids (vertical/subvertical and lateral access development excavations and stopes both filled and unfilled (including caving methods);
- Pillars and abutments of various dimensions and orientations.

These excavations and regions of intact rock interact in subtle and often complex ways that can be difficult to predict; necessitating the application of soundly based geotechnical engineering methods to ensure acceptable levels of mine safety, productivity and economic efficiency. The "design life", size, shape, orientation, rock characteristics and purpose of each particular engineering structure must be suitably taken into account when designing a mine. As would be expected, voids with longer design life and higher degrees of inherent risk require significantly more demanding design criteria. It follows that ore stoping / extraction methods selected by the mine operator will determine, to a large extent, each of the controlling factors mentioned above. These mining methods also greatly impact on the degree of exposure of mine personnel to hazards associated with those methods of mining. (For the purposes of this Code, stoping methods have been placed within two categories; entry and non-entry – see discussion below.)

Somewhat less obvious, but important components of MPD is extraction sequencing, production scheduling, and excavation methods (particularly where blasting is used) to develop mine openings throughout the mine. These components can significantly affect ground conditions in a number of ways (*e.g.* stand-up time, proximity to other workings, stress redistribution within and around the mine, and artificial damage to the rock mass). Further discussion on issues relevant to sequencing, production and excavation methods is provided later in this code. Consequently, MPD is to be an iterative process, whereby the "hazardous products" of each mining method are compared to expected ground conditions and required stabilisation techniques to determine the most cost effective, safe approach to mining an orebody.

Mine Voids

For the purposes of this code, the design life of mine voids has been divided into two categories;

- Long term voids; and
- Temporary voids;

with long term voids having a design life of at least one year. Examples of the two types of openings.

Table. Examples of Various Types of Long Term and Temporary Voids

Long Term Voids	Temporary Voids
Shafts	Drill drives
Drifts, adits, declines, headings.	Cut-off rises
Main dip and level development	
Escape ways and refuge bays	
Intake & return airways	
Offices and lunch rooms	Mill holes (drawpoints)
Workshops and Electrical substations	Extraction drives
Crusher and conveyor excavations	Working party magazines
Main pump stations	Stope ventilation rises,
Main magazines and fuel storage bays	Stope access drives

It can be seen that there are a number of voids that could be classed as long term openings; suggesting that a proportionate number of openings in a mine require rigorous and detailed design. The impact of each mine void on general mine stability can be further mitigated to varying extents; depending on GSR (Section 5) and the use of backfill.

Pillars and abutments

A pillar is an area of ore (in contact with both hangingwall / roof and footwall / floor materials) that is left to support the overlying rock, hanging wall or backfill. These structures are an integral component for the continued safe operating conditions in most underground mines. Pillars can be permanent - left in place for the life of the mine – or temporary - recovered sometime after their formation. Permanent pillars may also be stripped / reduced in size during the latter stages of a mine, when risk assessments allow. Pillars can be designed for mine-wide stability (*e.g.* intra-panel pillars) or work-place stability (*e.g.* rib pillars in an airleg stope). "Crush pillars" can be considered to be permanent pillars, as they continue to provide support to roof and floor strata.

Mine abutments can be considered as areas of unmined rock at the edges of a mine void that carry may large regional loads. Abutments can also represent an integral component of mine stability; generally as a zone of support for ground arching - say from a mine pillar, across an extraction panel, to the abutment. Similarly to the design approach for mine voids, the location, dimensions and orientation of mine pillars and abutments, the degree of detail, rigour and conservatism applied is based on risk assessment – should these structures fail or not provide adequate support as required by the initial design.

Issues to be considered when designing mine pillars include:

- Strength of materials – floor, pillar and roof – and their time dependent characteristics.
- Pillar cross-sectional area and aspect ratio.
- Variations (from pillar design) during to mining methods or spalling over time.
- Ground stress – natural, and stresses due to mine/pillar configuration or nearby mine voids.
- Orebody dip.
- Orebody depth.
- Geological defects.
- Shear strength at floor and roof contacts and influences of water at contacts.
- The use of GSR or backfill to improve pillar stability.
- The presence of important in-mine or other surface structures above the mine and the expected area of influence / angle of draw in the event of pillar failure.

Ore extraction mining methods

The MPD processes used for stoping / ore extraction in a mine are largely dependent on the intended mining method and whether mine personnel are required to re-enter the stope.

Entry mining methods

Entry mining methods (Hartman, 1992) include cut and fill, room and pillar, bench and fill, gallery stoping and shrinkage stoping. They have the common feature that the workforce is exposed to the potential hazard of rock falls from or collapse of large areas of stope backs and walls, particularly in wide orebodies. Entry mining methods typically require successive slices or lifts to be mined from the orebody, and hence require regular inspection and scaling and the installation of additional GSR for each lift. Therefore entry mining methods generally require a high level of effort in local-scale ground control; however this provides a benefit in greater control over such factors as ore grade, minimising dilution and maximising recovery as well as improving safety. Ground conditions have a very strong influence on mining method selection and hence they should be well understood before a commitment is made to the development of multiple stopes. The mining of a trial stope is obviously one approach that could be used in a new mine where there may be some concern about the suitability of the mining method for the expected ground conditions.

Non-entry mining methods

Non-entry mining methods include open stoping, sub-level caving, block caving and longwall and Wongawilli mining on retreat. Where orebody geometry and ground conditions permit there is a general trend to the more productive non-entry mining methods. These mining methods require a much higher level of technical input into large-scale ground control primarily because of the large dimensions of the area being stoped or caved.

The exposure of the workforce to potentially hazardous conditions is reduced where the work is conducted in development size openings that are usually supported during initial development of the stope. However, the potential exists for large-scale pillar and/or wall collapse in open stope mining methods. Such occurrences may cause rapid flow of materials into nearby mine openings, air-blast problems, dilution and blockage of drawpoints.

Conversely, caving methods may hang-up or not cave in a controlled manner and result in a number of hazardous conditions; including potential for massive airblasts. In fact, hang-ups and irregular caving, particularly at larger distances from the extraction level, should be expected to varying extents – due to the variable nature of most rock masses – and systems must be put in place to detect caving irregularities before they become problematical to the mine. The down-dip advance of the stope abutment can cause substantial adverse changes in the ground conditions, primarily because of increased mining induced stresses, to the possible detriment of existing or new development in these regions. Engineered fill is an integral part of the design and operation of some non-entry mining methods. Where fill is used, the design duty, transport, placement and quality control aspects of the fill system should be addressed in a systematic manner to ensure adequate and consistent fill performance during the life of the mine. More discussion of mine fill is provided below.

A balanced geotechnical stope / extraction panel design process should involve the integrated consideration of a range of issues including:

- Orebody geometry;
- Orebody quality and geological variability;
- Ground conditions in the backs / roof and walls;
- Rock fragmentation requirements; and
- Ground support and reinforcement requirements.

Mine fill

Backfilling in mines has been used over many decades - generally for one of three purposes:

1. Control of ground movements and stresses within the mining horizon
2. Control of ground movements above the mine horizon (mining subsidence)
3. Storage of waste materials.

The creation of large volumes of unfilled stope void can develop zones of high rock stress and exposure of significant geological structure to an extent that large-scale displacement (collapse or caving) occurs in an uncontrolled manner with little prior warning. Consequently, the inherent limitations of unfilled open stoping methods, particularly when combined with very high extraction ratios, poor quality ground conditions and a lack of geotechnical engineering need to be recognised by mine management.

There are limits to the wall and/or back areas that can be exposed before significant levels of dilution, crown pillar collapse, caving and/or mining induced seismicity occur. The very short term advantages that can be associated with unfilled open stoping methods need to be carefully weighed against the potential requirement to introduce a stope fill transport and placement system, often at short notice, to fill large unplanned voids. Where backfilling has been implemented as part of the general safe mining practice employed at a mine, there needs to be recognition that the mining process is not complete until the void has been filled within design limits with a suitable material.

Mine fill has a number of very important, but not widely understood, roles to play in large-scale ground control, including:

- Support of individual rock blocks on the surfaces of stope walls, pillar walls and backs (if tight filled) – preventing collapse of the immediate roof/hangingwall and also minimising mining subsidence;

- Lateral confinement of the rock mass, thus increasing its compressive strength;
- Provision of an adequate working surface in entry stoping methods;
- Enabling secondary extraction by the exposure of backfill is capable of standing safely during the extraction of the adjacent rockmass; and
- Damping the vibrational response of the rock mass during seismic events (Glen, 1993).

The provision of small amounts of lateral confinement to the rock mass, by the fill, can have a very beneficial influence in improving the strength of the rock mass. This fact is demonstrated in triaxial testing of rock core where a small confining pressure can increase the strength of the rock. Fill has the potential to play a significant role in large-scale ground control by providing support over very large areas of stope walls. The importance of the role of fill in large-scale ground control is often under-estimated. The systematic use of an appropriately engineered stope fill system in a mine can allow the mining of higher extraction ratios in a given orebody compared to a mine with no stope fill, assuming a non-caving method of mining is used. The improved safety conditions and higher extraction ratio are of direct benefit to the economic performance of the mine.

The most common forms of mine fill include cemented and uncemented waste rock fill, paste fill, hydraulic fill and sand fill. All fill types have their own set of hazards that need to be addressed during mine design, planning and scheduling, and during operations. Of all threats posed by mine fill in underground mines, the greatest threat would be from all fills reliant on water content (hydraulic and paste fills in particular). As mentioned previously, if not properly designed, implemented & monitored, these fills have significant potential to inundate large areas of the mine with catastrophic consequences.

Inadequacies in any of the following can result in failure of the backfill to act as an engineering / supportive structure and potentially inundation of the mine:

- Fill design (fill specifications must suit purpose).
- Barricade design (must suit purpose).
- Filling strategy (pour-rest cycles, curing time).
- Quality control at process plant.
- Barricade construction (*e.g.* problems with local ground conditions, excessive drive dimensions, poor quality of materials used, personnel training)
- Overall management systems (*e.g.* fill request process, quality and performance monitoring strategies, including trigger action plan for out of specification fill or damaged barricades, blasting adjacent to uncured barricades or fill, tight filling/ overpressurising).

- Mining through or adjacent to uncured/ unconsolidated fill, causing collapse or flow of fill into workings.
- Water level and fill level control (including recharge of drained fill in old stopes due to ingress of mine/ ground water). For example, water should not be allowed to accumulate in filled stopes, particularly those filled with uncemented sand fill. The accumulation of water in sand filled stopes can potentially result in the following:
  - Liquefaction of the fill by dynamic loading;
  - Hydraulic pressure on fill bulkheads or barricades; and
  - Hydraulic pressure on lined ore passes or ventilation rises in the fill.

The removal of crown pillars below filled stopes containing significant volumes of water should be treated with extreme caution and suitable measures to drain the water should be undertaken before the pillar is removed. Clearly, it is important that the mine operator manage paste and hydraulic fills with extreme caution. A suitably qualified person with expertise in the area of paste and/or hydraulic fill should always be used for fill & barricade designs and implementation of fill systems. Fill systems should be monitored closely by operations personnel and should be overseen and reviewed by suitably qualified and experienced personnel.

MINE DESIGN METHODS

Mine design methods generally in use by the mining industry include:

- Empirical methods based on precedented experience; and
- Numerical analysis methods.

In small scale mines with good ground conditions, it may be acceptable to use a combination of mining experience with one of the empirical design methods. In larger mines employing bulk mining methods, and in particular those with challenging ground conditions, it may be preferable to conduct a preliminary design using the empirical design methods in conjunction with one or more appropriate methods of numerical analysis, in conjunction with mining experience and stope monitoring techniques.

Empirical methods

The empirical design methods allow mining experience in a particular set of ground conditions to be incorporated into the stope design process. The general methodology of empirical design methods is that, several geotechnical parameters are assessed and given a rating on the basis of simple index tests or visual observations and tabulated numerical values. The rating value is then used to design the "allowable" dimension of mine structure, based on observations / performance monitoring of the same geotechnical parameters with similar ratings in other mining areas. For mine development and stope designs, these methods typically use one of the rock mass classification methods (Bieniawski, 1989), or a modified version thereof, to relate excavation geometry to the expected ground conditions. The stope geometry may be expressed in terms of "hydraulic radius" determined by dividing the exposed wall area by the wall perimeter. Several empirical design charts (*e.g.* Hoek etal, 1995, Hutchison and Diederichs, 1996) have been produced to aid the stope design process.

Similarly, several empirical pillar design methods are available in public domain – each having their own strengths and weakness in certain applications. Most are based on a mass-pillar strength, which is altered by a function of the pillar aspect ratio. In some situations, rules of thumb have been used, based on significant numbers of case data (*e.g.* acceptable pillar width to height minima for crown pillars, and the "indestructible squat pillar" theory for room and pillar-type structures), to derive pillar dimensions for mines. Prior to using empirical methods to design mine openings, pillars (and GSR - see later), the mine operator needs to be aware that these methods have their limitations (*e.g.* Palstrom and Broch, 2006, Milne and Pakalnis, 1997, MOSHAB, 1997, and DMP, 1999). Potential exists for the inherent variability of the rock mass to be obscured by the need to make it conform to an arbitrary set of tabulated numbers.

Regular visual observation of stope and pillar performance and, where practicable, instrumentation of mining structures, should be used to verify the continued use of any empirical design process. Where observed performance conflicts with empirical predictions, the mine operator must undertake one of the following actions; modify the empirical method accordingly; use alternative design methods; or derive any verify specific limits for the continued use of that method in that mine.

Numerical analysis methods

Numerical methods of stress analysis or block behaviour, in two or three dimensions, allow the interaction of nearby stopes and development openings to be considered in much more detail than is the case with empirical design methods. The use of numerical methods generally requires considerably more input data including the geometry of development openings, stopes and pillars, extraction sequence, rock stress field, rock mass properties and location and orientation of geological structure. The development of numerical methods has proceeded at a rapid rate during the past 10 years to the extent that there is currently available a wide range numerical modelling methods that can be run on most high-end personal computers.

The application of user-friendly numerical stress analysis codes may appear very straight forward at first sight. An appreciation of the challenges in: selecting the appropriate stress analysis code, having a sufficiently good model representation of the actual mine geometry being analysed (eg complex 3D geometry being poorly approximated by a 2D cross-section), the quality of the input data versus the inherent variability of the actual rock mass conditions, and having the mining experience and judgement required to correctly interpret the results in the context of actual underground observations; should all help to restrain the unthinking use of numerical stress analysis methods.

Considerable engineering knowledge and mining experience should be exercised to determine the appropriate numerical model for the task at hand; some numerical analysis methods have limited application due to their computational methods.

MINE ABANDONMENT

By the time of mine closure, there should be adequate data to address all the long-term geotechnical concerns in regard to the abandonment of a mine. Before underground mines can be legally abandoned, the Code requires that all long term drainage, environmental, and public safety/access issues are adequately considered and controlled. Environmental requirements for abandoned mines are specified by the license conditions imposed by the relevant Government authorities during the mining project approval process.

GROUND SUPPORT AND REINFORCEMENT (GSR)

The terms ground support and ground reinforcement are often used interchangeably, however they refer to two different approaches to stabilising rock (Stillborg, 1994). Ground support is applied to the perimeter of the excavation to limit movement of the rock mass. These methods typically require the rock mass to move on to the support to generate loads in the support. Ground reinforcement is installed beyond the mine perimeter, into the rock, to limit movement of the rock mass.

Ground support and reinforcement includes all the various methods and techniques of both kinds that may be used to improve the stability of the ground.

There is a wide range of rock reinforcement and support methods available that can be used individually or in combination to stabilise underground mine workings – including:

- Rock bolts, and cables;
- Friction rock stabilisers;
- Cast concrete lining;
- Shotcrete (plain and fibre reinforced);
- Thin flexible surface coatings;
- Strapping.
- Timber props;
- Timber packs;
- Timber or steel sets;
- Yielding sets;

- Hydraulic props;
- Screw-jack type props;
- Mesh; and
- Backfill (plain and strength enhanced).

A description of each method is beyond the scope of this Code, however a brief outline of some of the more commonly used surface rock support methods is provided below.

Mesh

The installation of mesh on the backs and side walls of an excavation is a technique that can largely eliminate falls of small rocks. Hoek, Kaiser and Bawden (1997) refer to this approach as a "safety" support system, the purpose of which is to prevent unexpected falls of small rocks. This type of support system, however, is not designed to support large static or dynamic loads. It can be used in conjunction with other elements to be part of an integrated system for supporting large dynamic or static loads. Providing that the mesh and the reinforcing elements do not become overloaded with loose rock fragments contained in the mesh, this system largely eliminates the need for regular check scaling.

These support systems may also be appropriate, irrespective of heading height, in regular travelways, places where the workforce regularly gather, and permanent installations containing specialised equipment or explosives; *e.g.* shafts, declines, main haulages, workshops, lunchrooms, offices, main pump stations, major electrical sub-stations, crushers, conveyors, main magazines, etc. There is a variety of mesh available. The three main types are welded wire mesh, chain link mesh and non-metallic mesh. Chain link mesh has greater flexibility than weld mesh, while weld mesh is better suited for use with shotcrete. Where corrosive conditions exist, galvanised or non-metallic mesh is recommended.

Strapping

Strapping is sometimes used to extend the area of coverage of an array of rock bolts. This method may be suitable for restraining relatively large blocks and assisting with roof beam development. Recent developments in wire-mesh-type straps can withstand reasonably large – scale movement and seismic events. Straps are usually installed in a linear or grid pattern along the side walls or the backs and can be installed in conjunction with other surface rock support. Its use as a control method is limited, as the rock surface between the straps remains largely unsupported and blocks still may fall out.

Shotcreting

There are two types of shotcrete application - wet mix and dry mix. Each method has its particular use in surface rock support. The current trend is to use fibre-reinforced shotcrete, referred to as fibrecrete. When designing a shotcrete program the following issues need to taken into consideration:

- Amount of shotcreting required;
- Shotcrete strength;
- Shotcrete thickness;
- Presence of groundwater (*e.g.* quantity, chemistry, pressure);
- Need for drainage of groundwater from behind the shotcrete;
- Water quality (potable);
- Type of shotcrete mix (wet or dry);
- Use of microsilica;
- Admixtures (plasticisers, etc);
- Accelerators (for wet mix);
- Fibre reinforcement;
- Curing (external or internal);
- Testing and monitoring;
- Correct spraying equipment; and
- Correct shotcrete application.

Other Surface Coatings

Recent developments include the use of synthetic rock surface coatings specifically designed for rock stabilisation in underground mines. Further work to develop these coatings may be required before they can be reliably and effectively used on a routine basis at minesites.

GSR Design

There are a number of ground control design methods that can be used. All these methods rely on having a good understanding of the prevailing ground conditions* before undertaking the design. Each method has its limitations and should only be applied to areas that are not impacted significantly by these limitations. The design methods that can be used include:

- Empirical or experience based methods developed from extensive local information [*e.g.* RMR (Bieniawski, 1973), Q (Barton et al, 1974), MRMR (Laubscher, 1990), US Army Corps of Engineers (Stillborg, 1994), Stability graph method (Hutchison et al, 1996)];
- Analytical/deterministic/limit equilibrium methods - using geotechnical parameters derived from either laboratory testing or back analysis of existing failures (*e.g.* composite beams (Peng 1978), support interaction analysis (Hoek et al, 1995);
- Kinematic/stereographic and block analysis methods (*e.g.* SAFEX; Thompson and Windsor, 1996), UNWEDGE; Carvalho et al, 1996);
- Numerical modelling (Hoek et al, 1995, Brady and Brown, 1993, Beck et al, 2010), and
- Physical modelling (*e.g.* Player et al, 2009).

- Seismic criteria (*e.g.* ejection velocity (Ortlepp, 1992), allowable displacement, rock damage criteria Kaiser et al, 1996).

Design criteria for each of these methods can differ; however, each design criterion is dependent on the level of acceptable risk of any particular failure and the degree of inherent uncertainty regarding the characteristics of the rock mass* and design method. It is recommended that a sensitivity assessment be carried out to determine the effect of critical geotechnical parameters involved with mine stability.

Any deficiencies that are highlighted in the analytical methods and sensitivity studies should encourage further work to remedy these matters, extend the use of the method or develop a new method. The most common forms of design analysis are the empirical and kinematic or deterministic methods for which there are several packages available commercially. These methods are relatively simple to follow. Numerical modelling allows the design of underground workings to be considered in much more detail than is the case with empirical or deterministic design methods. One of the drawbacks for the use of numerical methods is that they generally require considerably more data input, which cannot always be adequately provided or accurately measured.

Computer-based numerical modelling packages have developed rapidly during the past 10 to 20 years and this trend is likely to continue. A wide range of design packages are currently available that can be run on most standard mine site computers. However, it must be recognised that differences exist between the solution methods used by the major numerical modelling techniques - *e.g.* finite element, finite difference, boundary element, and distinct element codes. These different solution procedures can give rise to some variation in computation of stresses and strains. The design engineer must acknowledge the differences and limitations between each of the numerical model codes with respect to the problem at hand. The difficulty numerical model input data has in accurately representing the inherent variability of the complex rock mass* should also be recognised. It is therefore a prerequisite that significant mining experience and judgement is required to interpret and use the results correctly. It is also recommended that each numerical modelling technique be calibrated against observed ground response to mining for local conditions.

Considerable engineering judgement and mining experience is required to determine the appropriate levels and methods of geotechnical investigation required for the development of a geotechnical model of a particular mine, and to determine the method/s of analysis best suited for underground mine design. Geotechnical data for design can be obtained from a number of sources including: published literature, natural outcrops, existing surface and underground excavations, chip and diamond drilling (for determining rock mass strength*, structure, and hydrogeological data), geophysical interpretations, seismic records, pump tests, field tests, trial pits, and experience. It would be a statement of the obvious to say that the quality and usefulness of these sources of data varies widely. However, qualitative information is better than none and, if nothing else, such data can be used to identify the areas requiring more detailed investigation and analysis. The information gained from geotechnical investigations notably provides valuable information for mine design, but also assists with the development of a mineral resource estimate, and ultimately an ore reserve estimate. Particularly in marginal deposits, the geotechnical mine design limitations may define whether the resource can be classified as a reserve and therefore whether or not it should be mined. Again, all information used to establish effective ground control is to be maintained within the PHMP.

Obviously, the number, size, shape and orientation of the excavations and the ground conditions should be considered when selecting the most appropriate ground support and reinforcement system or systems. It is suggested that the mine operator ensure that GSR requirements are derived from a combination of these methods; using a sensitivity-type analysis, taking into account local conditions and experience. Any deficiencies that are highlighted in the use of these design methods should encourage further work to remedy these matters, extend the use of the method or develop a new design method. As stated in Section 4, prior to using empirical methods, the mine operator needs to be aware that these methods have limitations (*e.g.* Milne and Pakalnis, 1997, MOSHAB, 1997, and Palstrom and Broch, 2006); consequently a reasonable degree of conservatism is needed when relying on these methods.

It is recognised that during the earlier design stages there is usually limited detail of the overall rock mass* available, and that it is necessary to make a number of assumptions/simplifications to arrive at a balanced mine design. Rock support and reinforcement design methods must continue to evolve and develop as the mine matures. These methods, in keeping with the "engineering method", do not present an exact closed form solution with one unique answer. Rather, they are based on underlying scientific principles, strength of materials concepts, engineering computational modelling, static and dynamic testing plus considerable observations of field performance to present a range of solutions.

Corrosion is an important factor that needs to be considered in the design and selection of the rock support and reinforcement. The influence of corrosion will mean that virtually none of the conventional forms of rock support and reinforcement can be considered to last indefinitely; they all have a finite design life. The two main causes of corrosion are: oxidation of the steel elements, and galvanic consumption of iron by more noble (inert) metals, for example copper. Ranasooriya et al, 1996 and Villescusa et al, 2007 provide more discussion on the corrosion of GSR elements. The groundwater or artificially introduced mine water, for example hydraulic fill water, needs to be checked to determine if it has the potential to cause corrosion of the rock support and reinforcement.

With respect to GSR design, it is important that the mine operator recognise that ground conditions can change during mining due to a number of factors and that the effectiveness of GSR may also change with time. Factors contributing to change in ground conditions include:

- Loosening of the rock mass due to blast vibrations;
- Drainage of water from soft rock formations and jointed rock;
- Deterioration of some rocks on exposure to air or water over time;
- Sudden changes in stope geometry, *e.g.* formation of a cut-off slot;
- Stress reductions, or stress shadows, near large stope/extraction panels; and
- High stress levels in abutments and pillars.

These changing conditions require ongoing assessment of GSR design strategies and systematic performance monitoring (further discussion is provided in Section 6.1.2).

GSR Installation

The timing of the installation of ground support and reinforcement should be considered as an integral part of the design to limit the potential for ravelling of the rock mass. In those headings or workplaces requiring control, the delay in the installation of the ground support should be minimized as far as is reasonably practical. It is recognised that several hours may elapse from the firing of a development face, before the heading is clear of post-detonation explosive fumes, watered down, scaled and cleaned out ready for the installation of ground support and reinforcement.

Extended delays in the installation of GSR, in the order of weeks to months, may jeopardise the effectiveness of the ground control because of the rock mass loosening and consequent reduction in the shear strength that may occur. Significant changes can also occur in the rock stress field around stopes / extraction panels. These stress changes can have an important effect of the amount of radial confinement that is experienced by the rock reinforcement. A reduction in the confining stress, normal to a rock bolt, may adversely affect the load transfer capacity of the bolt (Hutchinson et al, 1996). When the ground conditions are sufficiently poor, the available time that the excavation will remain open and stable (the stand-up time) may be considerably less than the 24 hours mentioned above.

In these situations special measures may be required to promptly install ground support and reinforcement prior to the removal of broken rock from the face. Shotcrete applied to the exposed backs and walls, before the heading is cleaned out, is one approach that may be necessary or effective. Rapid placement of the ground support as soon as practicable after blasting, minimising the time that the ground has to stand unsupported, is likely to be important for successful mining in these ground conditions. "Workplace" ground support and reinforcement should be installed (and tensioned, if appropriate) preferably on a hole by hole basis or at the very minimum on a row by row basis, before advancing. If the ground conditions are considered to be sufficiently poor, or the potential for failure of a block is judged to be high, then a hole by hole installation technique must be used. The drilling of a large number of holes, prior to the installation of the ground support, is not considered to be an appropriate system of work in these areas.

GSR installed for the purpose of providing additional support (usually to account for expected large scale movements during the ore extraction stage) can be installed on a campaign basis with success. For safety reasons alone, it is vital to ensure the design GSR is installed appropriately. Economically, the importance of correctly installing GSR is highlighted when one considers the marginal costs associated with the different types of GSR in comparison to the fixed costs associated with hole drilling, installation and grouting (*e.g.* equipment depreciation, drilling consumables, transportation, grouting and labour etc.).

It should be recognised that GSR, together with their ancillary fittings, combine to form an overall ground support and reinforcement system that consists of different layers. Each layer has its own unique contribution to make to the success of the system. The GSR design method used should ensure that the appropriate elements of support and reinforcement are combined in such a manner as to produce an effective overall support and reinforcement system that is matched to the ground conditions for the design life of the excavation.Suppliers of GSR elements are encouraged to provide an appropriately detailed set of mechanical properties and instructions for the correct installation and testing for each element type. The end user of the GSR must (at least) be able to demonstrate that they are following the manufacturer's instructions for the correct installation of the equipment. Training courses and materials should be readily available at each minesite to ensure that the workforce is fully conversant with the type(s) of GSR in use. There needs to be a thorough understanding, by all those concerned, with GSR usage, their strengths and limitations. Further discussion on training is provided in Chapter 7.

The equipment used to install GSR should, where practicable, be purpose designed and built for the particular range of GSR elements in use at the mine. For example, the use of development drill jumbos in narrow openings, where the GSR cannot be installed according to the design standards, is not considered good practice. Where the use of purpose-built installation equipment is not practicable, the mine operator must formally develop and implement appropriate GSR design and work/installation procedures to minimise risk to the workforce when installing GSR elements.

Examples of issues to be considered in the formal design and installation procedures include:

- Knowledge of ground conditions in the area where the GSR elements are to be installed
- Timing of the reinforcement installation - taking account of the potential for early deterioration of ground conditions and the ability of the reinforcement to contain this
- Scaling requirements - prior, during and following the installation work
- Reach and capacity of the equipment - needs to be matched to the opening dimensions
- Placement of the GSR element(s), including mesh, on the equipment prior to installation to be carried out from a secure position
- Correct alignment of the support or reinforcing element relative to the orientation and shape of the excavation and local geology
- Appropriate operation of the insertion device - *e.g.* if a drifter is being used, the mode of drifter operation should be "percussion off" or "no percussion" while travelling up the slide
- When to use rotation only (no percussion) when installing rock bolts using a jumbo – *e.g.* no percussion when tensioning threaded reinforcement elements
- The required torque to be applied to a rock bolt or dowel nut etc. without damaging the individual components, and
- Personnel movement during the installation of GSR.

GSR Quality control

The importance of quality control to the successful design and installation of an adequate ground support and reinforcement system needs to be clearly recognised and proper quality control procedures put in place. The supplier of the rock support and reinforcement system elements should provide information on the factors that determine the quality of the installation.

The mine operator should formally develop and implement appropriate GSR quality control systems / procedures to ensure GSR design standards are being met. When designing a quality system the following issues (not excluding others) need to taken into consideration:

- Storage and handling strategies for GSR elements must effectively minimise damage and deterioration to the elements
- The suitability of GSR elements in variable ground conditions – *e.g.* expansion shell bolts are generally ineffective in soft rock - as are friction bolts where correct hole diameter cannot be maintained
- Correct hole length is drilled and holes are flushed clean of all drilling sludge
- Orientation of the hole is appropriate for the excavation geometry

and expected block movement - axial tensile loading of the steel elements installed in the rock is generally preferred; shear loading should be avoided

- Hole should be drilled nearly perpendicular to the rock surface - use of hemispherical ball and domed plates may be required where this cannot be achieved
- Load capacity of the anchorage method, bar or tendon and surface restraint fittings should be appropriately matched to prevent the premature failure of any one component
- All steel and other components designed to be encapsulated in resin or cement grout should be clean of all oil, grease, fill, loose or flaking rust and any other materials deleterious to the grout
- Where full grout encapsulation of the steel elements is required, the method of grouting should show a grout return at the collar of the hole; other methods that can demonstrate complete hole filling may also be appropriate
- Correct tensioning or loading procedures should be used for the various rock support and reinforcement systems
- Plates and/or straps against the rock surface should have the required thickness to prevent nuts or barrel and wedge anchors being pulled through the plate and/or strap at the ultimate tensile strength of the tendon when loaded against the rock surrounding the bore hole
- Corrosion issues are recognised and remedied
- Blast vibrations may loosen threaded reinforcement systems
- Load tests are regularly carried out on point anchored rock bolts and friction anchored rock bolts
- Fully grouted reinforcement systems should be checked on a regular basis to ensure that the grout strength and encapsulated length of the bar or tendon is adequate
- Implement an action plan when it is found that the load capacity of the installed support or reinforcement system, grout strength and/or encapsulated length does not meet specifications
- Storage of resin grouts should be at the temperature range recommended by manufacturer
- Resin grouts are consumed before their "use by" date, or within a specified period of time
- Mixing of resin grouts should be for the recommended time and at the recommended speed - these should not be exceeded
- Cement grout is mixed at the recommended water: cement ratio, at the recommended angular speed in the specified equipment for the required time
- Water used for cement grout mixing is of the required quality or

the cement used should be able to develop the required uniaxial compressive strength with the run of mine water supply

- Any additives (*e.g.* retarders, accelerators, fluidizers, etc) to the cement grout mix are added in the recommended amounts and at the specified time in the mixing and pumping process
- All grout mixing and pumping equipment to be cleaned and maintained on a regular basis
- Any pumping equipment used to pressurise rock support and reinforcement should be regularly maintained and operate at the recommended pressure
- Shotcrete mix specification should state the slump of the mix, the uniaxial compressive strength and a measure of the toughness of the product at specified time intervals prior to or following mine application, as appropriate
- Samples of the mine shotcrete mix should be collected at specified intervals, under normal mine operating conditions, and tested in a NATA registered concrete testing laboratory for compliance with the shotcrete design specifications, and

Shotcrete thickness should be tested regularly during placement to ensure that the specified thickness has been applied - a means of permanently marking the shotcrete surface with a depth gauge probe may be appropriate.

Index

A

Abandoned Mines 37, 250, 251, 291
Acid Rock 51
Air Core Drilling 160, 163, 164, 166
Air-lift 139
Amalgam 209, 210, 211, 212
Aracoma Disaster 241

B

Backhoe 112, 120, 140, 141
Beam Engine 151, 152, 153
Bed Leveler 141
Blasting Machine 147
Bucket Dredging 139
Bucket-wheel excavator 27, 54, 130, 131, 134, 136, 137, 142
Bucyrus Model 114, 115
Bulk Mining 73, 289
Bullion Production 175

C

Calcining 199, 211, 212, 213
Catalogue Number 6, 12, 13
Coal Mining 19, 26, 29, 32, 49, 51, 55, 56, 61, 62, 63, 65, 75, 76, 77, 81, 83, 121, 123, 128, 131, 240, 241, 242, 243, 244, 245, 246, 247, 248, 249, 250, 251, 252, 253
Contemporary Usage 91
Cupellation 201, 202, 203, 204

D

Deepest Mines 74
Dragline Excavator 123, 124, 125, 126
Drift Mining 27, 76, 77
Drilling Method 158, 161
Drywasher 196
Dynamite 75, 144, 145, 146, 147

E

Electrometallurgy 174, 197, 199
Empirical Methods 289, 290, 295
Emplacement Areas 17
Excavator 27, 40, 54, 62, 110, 118, 119, 120, 121, 122, 123, 124, 125, 126, 130, 131, 132, 133, 134, 140, 143, 173

F

Fire Refining 99, 101
Friction Bolts 70, 298
Friction Hoist 149, 150
Froth Flotation 94, 96, 97, 179, 183, 184, 185, 186, 187

G

Giant Stripping 118
Gold Panning 192, 193
Grinding 27, 28, 94, 173, 174, 175, 177, 186, 187, 188, 189, 190, 197
Grouted Bolts 70, 71
GSR Design 296, 297, 298

H

Heap Leaching 60, 96, 134, 205
Heavy Metals 51, 60, 143, 144, 145
Highwall Mining 56, 57
Hydraulic Mining 21, 45, 48, 86, 87, 88, 89, 90, 91, 92
Hydrometallurgy 174, 197, 198, 199, 204, 206, 207

I

Ion Exchange 206

L

Land Surveying 213, 219, 221, 222, 223
Landfill Mining 26, 39, 40, 133

M

Metal Chelation 206
Metal Refining 174
Mine Fill 187, 287, 288

O

Operational Flow 41
Ore Access 69
Ore Removal 74

P

Personal Fatigue 251, 252, 253, 256, 258
Pipe Used 158, 171
Policy Framework 232
Prehistoric Mining 20
Pyrometallurgy 174, 197, 198, 199, 204

R

Rig Equipment 169
Rock Strata 268

S

Smelting 23, 24, 93, 95, 97, 98, 99, 100, 198, 199, 200, 201, 204
Snagboat 141
Solvent Extraction 95, 96, 198, 206
Steam Shovel 109, 110, 111, 112, 113, 115, 116
Suction Dredger 137, 138
Surface Mining 21, 25, 26, 27, 28, 35, 47, 48, 53, 56, 57, 58, 61, 63, 64, 66, 75, 103, 124, 130, 132